Home-Made Europe

Home-Made Europe

CONTEMPORARY FOLK ARTIFACTS

VLADIMIR ARKHIPOV

FUEL

Foreword

Jeremy Deller

Shack. Dungeness, England. 2004

I have a mild phobia of Do-It-Yourself and its attendant culture, I get anxious going to those large shops that sell drill bits and paint. Maybe it's the way it has become a hobby that seems so pleased with itself that makes it such a turn-off. I get the opposite feeling looking at these objects of resourcefulness and wit, which have a healthy disdain for traditional DIY values of finish and professionalism. They are more crafty than craft, concerned with getting by – and in some cases sticking two fingers up at the situation that necessitated these improvisations.

Artists, myself included, are excited by this kind of production where the skill is in the leap of imagination to transform one object into another. This transformation is of course an artistic process, and looking at this book it's easy to get a sense of the thrill of this alchemy where two plus two equals five. Some of these objects are so raw they still have the energy of their creator buzzing around them.

A note of warning though: to our rather pampered eyes many of these objects look like art, but in actual fact art looks *like*, if not aspires to be like, these objects. We owe a lot to this practical folk culture.

CONTENTS

Rei Kerçova Tirana, Albania. 2005

I'm cooking sweetcorn. You have to cut it in half because it's heavy. If you grill it whole it will fall and snuff these out! Now we have to wait a little until it's grilled... I like cooking it myself, and it saves money. I made it two days ago, but I'm going to add more wire to grill it better, and make it closer to the flame. It doesn't taste so good, it tastes like it's not completely cooked. These marks are caused by the smoke. But once it's grilled for a while, you can wash it and remove the smoke marks.

Wood, wire, glue, candles

Adem Bukri   Tirana, Albania. 2000

I'm a Llustra-Shoe cleaner. I've worked cleaning shoes for years, I've been in the cooperative since 1974. We make all the chairs ourselves. This is constructed completely from wooden slats. They're all hand-made, we are artists, craftsmen. Each shoe cleaner makes his own tools. If the carpenter made it he would charge 20,000 lek, and we make them for 10,000 lek. We can make it for this price because we find all the material ourselves, we work together as friends. We are born for this job, this is what we understand – other things we don't understand! This chair is for the client, it's higher than my chair. I shine your shoes, I am seated here, on the chair of the worker. Your seat is there. These are the rules. At night I have to put the chair inside the building. I leave my tools and equipment in another place. If I leave it here it will fly away, somebody will steal it.

Wooden slats, nails, carpet

LADDER

Guiljelm Paci   Tirana, Albania. 1974

Thirty years ago or more, we made this ladder. It's made by master craftsmen. We had eleven vines, to give us shade. Then I realised there was too much shade in the house, we needed to cut the leaves from the vines. So I cleared them all using this ladder. I bought up a lot of cheap scrap iron and I took it to a blacksmith. At the time they didn't sell ladders in the shops. We could buy one made from wood, but they didn't sell any made from iron. And it was cheaper. I've kept it ever since. I mean, it's still useful for other things.

Iron

DIVING MASK

Ferid Kola   Shkodër, Albania. 1978

When we were students, it was the period of the famous revolutionary triangle: education, work, and happiness. One of the most important classes in the school was work experience. One of these work placements took place in the hills of Lukova on the Ionian coast, where the country's youth were digging terraces in the hillsides. It's a beautiful place – artists were drawn there to develop their talents, to catch the essence of life, to be in contact with it. This was not the only work we did; we also worked in factories. We did some work experience in a factory where we melted copper: it was full of smoke…

Since I was a boy my hobby was fishing. I used to fish a lot on the coast, when we were students, the clear water intrigued me and I wanted to go diving. I was really interested to see what was down there. A friend from Lukova had a mask and a spear gun. So this was how I first began to explore the sea, and I fell in love with it. It was 1974. When we came back to Shkodër, myself and an artist friend of mine, thought we'd make these things ourselves, because they didn't exist here then, there were no spear guns or masks. So we started to make them, helped by our friends who used to work in factories. This is not the first we made; it's the fourth or the fifth. One sank in the water, one broke, one was lost. This one is from 1978 or 1979. This was the best I had, because I made it properly to fit my head… It was before I made the spear gun, because this is older. First we made it using clips, then elastic, then fasteners – we made so many and kept improving them.

Aluminium, perspex, rubber

Sokol Kola   Shkodër, Albania. 2005

As recounted by his father Ferid: He made the wooden bench so he could lie down to lift the weights. It was a drawing stool before, you sit down and fix the paper in front of you and draw, like this. The bar is a section from the axle of a car, and the dumbbells are wheels, which they took from the goods-wagon of a train. At the time he made it, there was a lot of scrap around. He wanted to work out, and he saw he could use the scrap to make the weights. He always wanted weights. When he was young he loved karate, now: bodybuilding. With this, he saw the pieces, and thought about how he could make the weights.

Wood, car axle, metal train wheels

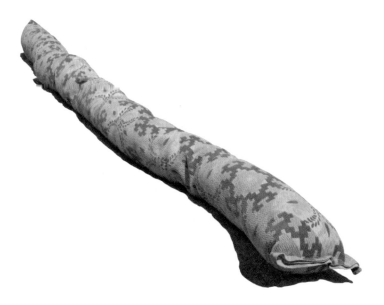

Brixhida Paci   Shkodër, Albania. 2002

This was made just to protect the room from the cold, from the wind during the winter. This stops the draught, but at the same time it even prevents the rain from flowing inside. It's a draught-excluder, but it's more like a barrier. I put it by the door, because there is a little gap between the floor and the bottom of the door. It's not traditional. It's just something I thought of myself. What's the best way of putting it, let's call it a creation. How can I stop the wind? How can I stop the rain falling in the window? How can I stop it flowing down the walls? There is wool inside. I sewed this one into a tube form, and then I stuffed it until it became the shape it is now. Then I sewed it up.

Fabric, wool

Agron Buzali Tirana, Albania. 2003

It was already made, the legs and the back. No wait, I fitted the back, and the seat as well. I strengthened it with this metal rod around the bottom edge. I work now, so I have enough money to change it. I would have burned it, but a guy asked if I would give it to him...

Chair, reclaimed wood, steel

Agron Buzali   Tirana, Albania. 2001

What can I say? It's a wooden shelf. What can I say? We display the goods going up the 'stairs', to the top. It makes them stand out more. I got the material from dustbins. I took it and sawed it up by hand. I measured most of it, but some parts I did by eye.

Reclaimed plywood, linoleum, nails

Adrian Gora   Tirana, Albania. 2003

To start with we mounted this kind of axle. Then we welded the forks, we call these parts the forks. They move, like this. Then we mounted the back part. On this thick piece we put the wheelbarrow. Then we put the wheels on. That's all: simple. And this part, at the back is a bicycle section. We sawed the bicycle in half and fitted the back end of it here. That's how we make it move. That's all, it's not so difficult... This wheel at the back is from a motorcycle. We put the brakes here, and these are the bearings that make the barrow move. We steer it with the help of this moving section with bearings underneath the barrow. That's all... We can also mount it with a motor if you want. You can put different motors in, motors in that have gears for speed. We cut it to put them in. We take these shoes, which we buy in a shop at a low price, and we sell them. We work all over Tirana, earning money to survive. With this we don't need to pay taxes, so I use it to ride around the city. They are made by us, by our people: Egyptians.

Bicycle parts, motorcycle wheel, steel

Vangjel Lashi   Tirana, Albania. 2005

We use this when we park the car, otherwise someone will take our space. I made it especially to park my car, but also for the cars of the other shop owners here. People come to the club in front of us, and they park their cars here. I thought this would stop them. When I went abroad, to Italy and Greece, I saw these constructions. They make something to save their parking spaces, they might use other materials, a plank, or something. I thought I'd make this. It was nice before, easy to park. But now it's crowded up and down in the street. I'm always waiting for the car to come, I load and unload here all day long, for my work. I don't have time to order a parking sign. It's not built to last, when I've finished with it I'll just get rid of it.

Wood, nails

Artur Kocia   Shkodër, Albania. 2005

This is a pair of boots that we use when we are working here in the winter.
We cut the tops off with scissors, make holes like this, and it's all done. Most
fishermen have boots like this, because now Shiroka is full of restaurants and
bars. You know how it is in these places: they throw lots of bottles onto the
floor. They break, and if you step on them you'll get cut. We got cut sometimes
just swimming in the lake. The sandals they sell in the shops don't fit properly.
And because they are low quality – fake leather – if you wear them in the water
once, they get damaged. But these will last the whole summer season, maybe
even the next summer, if you don't chuck them. They are strong. The most
important thing is that you are reusing your full length winter boots, that's
why we cut them down in the summer, so we can still wear them.

Soviet rubber boots

Harald Holnsteiner   Aigen, Austria. 1999

We made this just for fun, so we could get around the meadow. It doesn't really serve any purpose. We had to re-install the battery and the fuel tank, as well as all the wiring. The seats were newly installed too. We made struts for the steering wheel, overhauled the air-brake system pipes, repaired the speedometer, and tightened and reinfored the panels. We removed the body of the car to reduce its weight, and because we didn't need it. We'd already made the chassis back in 1994.

Car parts

Peter Sachsenhofer   Ulrichsberg, Austria. 2005

I made this hoop for basketball training. My coach thought I should put in some extra training because I wasn't quite good enough, so I made this. A shop-bought one would have been too expensive for me – about 250 euros. I wanted to save some money, so I built this one three weeks ago, over Easter. It's a bit lower than a normal basketball stand, because I couldn't find a taller pole. The bucket was used to store paint before.

Plastic paint tub, wood, hardboard

Ludwig Hehenberg   Haslach, Austria. 2003

I made this bread paddle three years ago. As far as I know you can't buy one like it. My wife bakes bread and she slides the dough into the oven with the paddle. Once it's baked, she uses it to take the bread out. We only eat the bread we bake, it has more vitamins, and is more filling. It's healthier and even tastier!

Wood

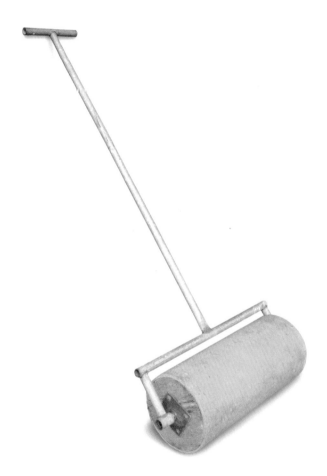

Ludwig Hehenberg   Haslach, Austria. 1984

This roller is used to roll in lawn seeds. I made it about twenty years ago. It has been used a lot, not just by me, but also by many of my friends. This is made from a sewage pipe that is filled with sand, with pieces of wood at each end. The rest of it is made from water pipes that have all been welded together. The roller has to be heavy so that it flattens and compresses the lawn. Because I studied metalwork, I was able to put it all together myself. I've never seen rollers like this for sale.

Concrete, metal pipe, wood, screws

Willibald Pfleger   Aigen, Austria. 1995

I built this merry-go-round ten years ago. I found the wheels at the radio broadcasting centre in Alteisen, where my son works. It really gets used when the children are here; the grandchildren and the kids from the neighbourhood. The armchairs were once used in a show at the television centre. They would have been thrown away if I hadn't used them.

Wooden cable roller, seats, metal, tap screw

Martin Bogner   Aigen, Austria. 2002

I built this go-kart with my dad last autumn. We found the scrap material at granny's house, she'd thrown it away. The front wheels come from a skateboard. Dad cut it and drilled a hole in it, and put in a small rod. The seat is attached with a wire strap. You use this piece of wood to steer with your feet, and this bit here is a wooden brake. We usually attach it to a bike.

Skateboard, barrow, wood, bolts, rope

Rudolf Berger   Aigen, Austria. 2004

I made this two years ago, when we had lots of concreting to be done. The soft concrete is poured into the hole and you use jerky movements with this tool to level it all off.

Wood, nails

Arnold Berger Aigen, Austria. 1998

As recounted by his father Rudolf: When we built this house, here in the Ulrichsberg district, we could choose between keeping the old dustbin or buying a new one. We kept the old one as we still had debts. But then the bottom of it rusted through, so we re-riveted it. My son attached the wheels because it's quite far to the street along our drive. Previously, it had taken two of us to take the bin out. Later on, they introduced plastic bins and we didn't need this one any more. But we kept it, because we liked the wheels – they'd come from a pram that we'd thrown out. It was quite a bit of work, the iron sheet had to be cut and riveted.

Dustbin, iron plate, pram wheels

Johann Brandl   Aigen, Austria. 2006

I made this last winter. Of course, I could have bought a similar one in a shop, but when you make it yourself, it's cheaper and as well as being a good pastime. Otherwise what would I do?

Wood, screws, metal brackets

Rupert Holnsteiner   Aigen, Austria. 2003

I made this frame with my son three years ago. It was just a piece of steel pipe before, then we welded a cap on the end of it. We only made this because you can't get as high a temperature using the ones from the shop. Because it's so big, you can fit a lot of people around it. We use it about five times a year. We move it to exactly where it's needed. For example: in the village when we put up the maypole; for barbecue parties; at midsummer; and so on. We've also used it in winter for a birthday party.

Steel

Herbert Gabriel   Oepping, Austria. 1985

I made this sledge ten years ago. At that time we didn't have a horse, but my brother-in-law had horses and I built the sledge for him. When we got a horse ourselves I got the sledge back from him so I could use it. The cover protects you from piled up snow and lumps of ice, which sometimes fly up from the horses' hooves. It only offers very basic protection. At first, instead of the cover, I used Plexiglass, but this broke – so I put this cover in instead. I made it quickly and very cheaply, I just thought up the design myself.

Metal, wood, plastic sheeting, cloth

Herbert Gabriel   Oepping, Austria. 1995

Wood

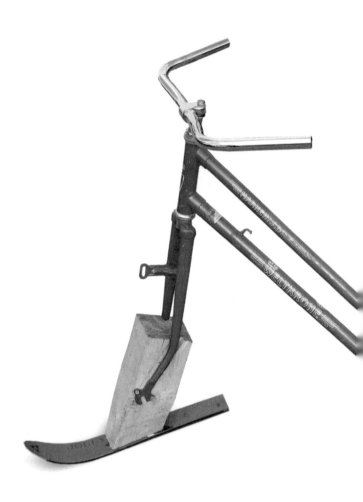

Creator unknown

Aigen, Austria

Bicycle frame, wood, skis, screws

Mag Gerald Bogner   Julbach, Austria. 1989

This pan is made out of rust-free, stainless steel. The upper part was made by machine and the bottom welded onto it. The handle is from pear tree wood, and treated with flax oil from an oil mill in Haslach. It's mainly used to make scrambled eggs. We have a bigger one, which is used for risotto. I made this myself because good quality pans are very expensive, at the time I was a student with no money. Besides, I had all the necessary materials at home, and I wanted to design the shape of the pan myself. I wasn't just thinking about its function, I wanted a good-looking kitchen too.

Stainless steel, wood

Franz Poltl   Aigen, Austria. 1997

This is a hay wagon that I made when the children were still small. We often went out walking and when they got tired, they would always sit in the cart. I made a canvas cover for hot days, which gave them some protection from the sun. You couldn't buy a hay wagon with space for three children. The ones for sale were all shorter, and could only fit two. That's why I built this one myself. I found the material for the canvas cover at my parents-in-law's, it's really old. We stopped using the wagon when the kids got too big for it. Now it stays in the attic.

Wood, blanket, wheels, screws

Franz Maurer   Klaffer, Austria. 2000

It's a good five years since I made these. You can buy ones like them, but they're too expensive. They stop you from sinking too deep into the snow. I saw a similar contraption somewhere, so I built these using the same design.

Wood, leather, screws, rope

Fritz Sonnleitner   Aigen, Austria. 2002

This is a birdhouse for titmice. I made it in the wintertime two years ago. I made several like it, because I enjoy doing things like that. I hang them in the fruit trees around the garden.

Wood, nails, wire

Willibald Pfleger   Aigen, Austria. 1980

This was a wheelbarrow that was used to take out the rubbish. I modified it with iron wheels from an old threshing machine, so that I didn't have to think about inflating any tyres. When I need it, I simply take it and go. The frame is made from a car bumper, the body from aluminium alloy. I bought that for a good price in those days. The wheelbarrows that you can buy in a shop don't have such a high load carrying capability. This one can hold heavier loads and you can fit more into it. A normal wheelbarrow doesn't last thirty years, like this one has – it would have broken a long time ago.

Metal sheet, bolts, metal wheels, metal pipe

Hilda Leitner   Haslach, Austria. 2002

I had to make holes in this, to let the water pour out from the bottom. I wrapped it up securely so that it didn't fall apart. I used this one for four summers, although I don't put plants in it any more. I have many more of these buckets in better condition. I've wanted to throw this one away for a long time, but it can still be used as a rubbish bin.

Bucket, tape

Franz Leitner   Sprinzenstein, Austria. 1990

This was built in early 1990. It's for hunting, to allow me to stay hidden and observe the game, to have a roof over my head when it rained, and also for shooting of course. I put padding in the openings, to get a clean, quiet, and targeted shot: to prevent the animal suffering. Each hunter has to check the stability and usability of the hides in his area at least once a year. Rickety or rotten rungs have to be changed. The load capacity of a ladder must be up to 150 kg, and it must also be standing at the correct angle, to be used safely.

Wood, nails

Mr Kajbr   Prague, Czech Republic. 1995

It started about six months ago, when I didn't say hello to a woman who lives in the block – I had my reasons. I was about five metres from her, so we looked each other in the eye. I had no idea it was going to end up like this. After about five minutes she ran down the stairs and locked the door on me – just like that! She locked it right in front of me... Then I thought to myself, what if she's going to lock the door on me again? So I took my keys and as I was coming back I actually saw her running down and locking the door again! That's why I had to climb in through the window... It's hard to explain. This sort of thing has been going on for years, since the day I moved here. The whole neighbourhood is rather odd. When I first arrived an old woman who used to live at number seven actually said to me, 'Oh, Mr Kajbr, this isn't a nice neighbourhood to live in at all – you won't be happy here.' And really, she was quite right. I haven't experienced any joy whatsoever while living here. It's just not right.

It wasn't always like this, I was lucky when I lived in the rest home in Moravia. I had support from some influential people at that time, and had been like that ever since my time at the academic institute. I wouldn't call it good connections, but you could say it was back-up of some sort. However people were obviously misunderstanding these circumstances. It was all simply because I knew my craft really well, I was just doing my job well... For example while I was in Moravia, I received a number of letters from the prime minister, but people refused to see that it was only for my work, nothing else...

So now I'm using this little step to climb in through the window. At first, because the log was broken in half, I just used it to chop wood on. Then I used it to rest my elbow on while I was sitting. Later I put a screw in it, because it was wobbly, and then I used it to rest my leg on when I was studying the Bible. I would lay my leg on it and put the books on that leg. It was only later I used it outside. I've put it next to that window because it was too heavy for somebody to steal it. I don't think anybody would be able to take it away with them.

Wood, wire, screws

Martin   Prague, Czech Republic. 1991

I made this little thing about ten years ago when I was using an old camera. It didn't have a lens cap for the lens, and it was impossible to buy this particular cap in the shops. So I decided to make one myself. My intention was to make the sort of cover that would stop light shining onto the film, and that I could use instead of a shutter. That's why it's black and lined with velvet inside, so it doesn't let in any unwanted light at all.

Plastic tub, velvet, tape, glue

Nev Richards   Bishops Castle, England. 1997

It's a yeast paddle, for skimming yeast off the beer. It takes the yeast off and the beer drains back through the holes. That's the lid from a plastic bucket, and the handle is just a little piece of scrap stainless steel pipe. I made it four years ago, before we found something a little bit more hygienic... Oh, everybody does their own thing. A very popular tool for doing the same job is a frying pan with holes drilled in it, believe it or not. I've been working most of my life in engineering. When I became redundant a few years ago I decided to start brewing, and basically here we are.

Plastic lid, stainless steel pipe, screws

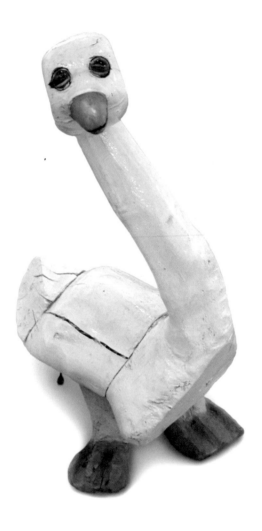

Sidney Pearce

Shrewsbury, England. 1955

Carved wood, paint

Mike Bate   Shropshire, England. c.1970

It must be about thirty years ago now I moved to Haroldstone. I had a lawn, but no roller, so I decided that I'd make one. I got an old milk churn, put a bar through it and filled it up with concrete. Then I made the frame with materials that I had lying around. I used it to roll my grass as it grew. This is the story of the roller. As you see, I can still use it now, it still works. To buy a roller thirty years ago would have been rather expensive. I simply hadn't the money, and why buy something if you can make it? But it's just the way things were then: you used the materials you had to hand. More often I would use it in the course of my work – doing a building job somewhere, converting a house, when we create a new lawn. It's used for that purpose more often. I've been a builder for, well, thirty years. And I spend quite a lot of time playing around with things.

Milk churn, concrete, scrap metal

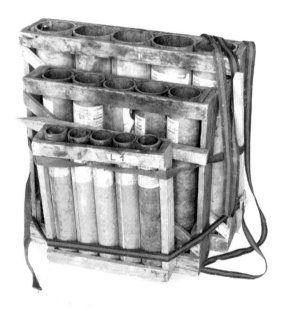

Can-Can   Le Tourne, Bordeaux, France. 2008

As recounted by Marie-Claude, pictured: It was made by someone from
Aveyron last year, for the wooden boat festival. You have to put some gunpowder
in the tubes. After that, it's electric, they're connected... And they fire.
Everything is hand-made, of course! They took large cardboard rolls, the ones
that are used for carpets, and sawed them up. Then they put them inside the
frame and fixed them in. And away you go! You can attach it to a boat if it doesn't
burn, or to a post near the site, or in the trees... There are electric detonator
wires – one, two, three, four, five – so I light this one and it's number one that
goes off, or the two, or all of them. We use it a lot.

Wood, cardboard tubes, ribbons, nails

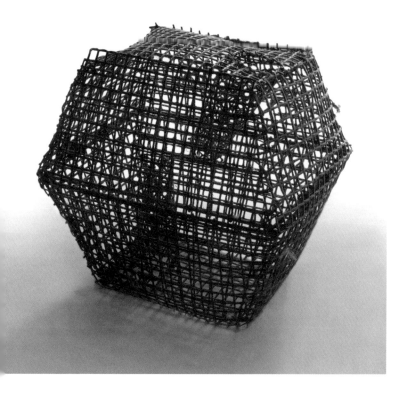

Julien   Le Tourne, Bordeaux, France. 1990

As recounted by Marie-Claude: It's for fishing. You open it and put rotten fish inside, then you place it at the bottom of the river with a dead fish floating on top. You come back one or two days later, and eels have gone in through the hole. They eat the head of the rotten fish, then they can't get out again. They can't get back out through the hole, they can't find it, so they're trapped. Then you open it, take the eels out and eat them. We still use it, there are more over there. You can make them with woven bulrushes too. Jojo made this one, he used to be a fisherman. He was an oil tanker captain, but he knows all this stuff really well. He has a boat and goes out on the river eel fishing.

Plastic net, wire

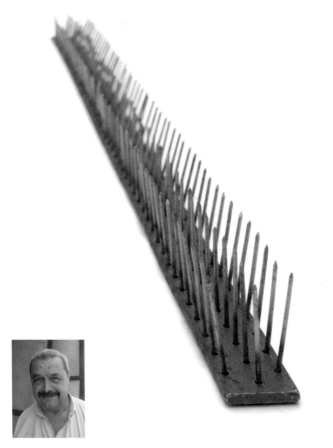

Florian Brun

Bordeaux, France. 2007

As recounted by Lyonnel Aeck, pictured: This was made by Florian Brun, the new blacksmith. But we're not allowed to use it any more, because it's forbidden. Now they use the same system, but with plastic sticks, so that pigeons can't sit on them. That's what it was made for. It was an early version. We always make small things, and then when we don't use them any more, we leave them and make other things… Usually they're made to be put on the ledges of buildings. Like on that house opposite. Pigeons sit there almost every day and it gets covered with bird droppings. Pretty birds, those pigeons, but they make everything dirty.

Metal, nails

Florian Brun   Bordeaux, France. 2006

As recounted by Lyonnel Aeck: This is an iron mould used to cast bollards that
are placed in alleys or pavement intersections, to stop vehicles from parking
there. You see – like over there? That's a low wall, but usually it would be a
post, a large octagonal post. This is a mould for making large bollards, there's
a different system for the smaller ones… With this metal form we tried to make
a mould that would cast posts that were identical to the original ones, we tried
to respect their shape, more or less.

Metal sheets, bolts

Creator unknown   Bordeaux, France. 2006

As recounted by Lyonnel Aeck: It's for when they come to collect the rubbish, to level it out. We use the three prongs to flatten everything. When they pour the rubbish out of the bins, it falls into heaps, so we level it. That's it. Simple. It's the blacksmith who made it. Not the new one, the one who retired two or three years ago. He took a piece of pipe and then welded on three pieces of flat iron, which he curved and sharpened at the ends. Every time the dustmen came, they would empty the bins and there would be mounds left, and we didn't have anything to level it out. Otherwise we had to climb in there and do it by hand. So we tried to find something practical, simple, and cheap. It was made because we care about tidiness. So there you go! It's all these small tools that we didn't used to have.

Metal

Didier Poullier   Bordeaux, France. 1994

This is a crowbar for lifting large crates. We used to slide it under the crate and then lever it up so that you can put chocks or trolleys underneath. My colleagues and I made it a long time ago, at least fifteen years back. We had to have something that we could manoeuvre really well, and you can't find these in shops. It works using the same system as you use with a small crowbar when you open crates. These are old castor wheels that used to run on the rails around our stockroom. We made everything, welded it all. It was only used for moving crates. But we don't use it any more, now that we have all the machines. Before, we had to find a way to manage without lifting everything by hand, without bending our backs all the time.

Metal, castors

Dominique Biais   Caychac, Bordeaux, France. 2009

I made this by attaching a motor from a fridge to an old compressor tank that I had in my dental surgery. The motor was finished, so I kind of salvaged it. The tank contains air, there is a regulator, and the fridge motor. I use it to blow up my tyres, or as a spray gun.

There is a coupling at the end of the pipe, you screw it here, plug it in, and away you go. This part compresses the air that's held in the tank. And the manometer allows you to limit the pressure. When you reach 8 kg of pressure it stops, thanks to this guage here.

Steel cylinder, compressor, manometer, wheels

Rolland Van De Rhe

Bordeaux, France. 1985

Wood

Creator unknown   Bordeaux, France. 1979

As recounted by Nicolas Dickets, pictured: I can't remember who made it. It must have been someone who came by here maybe some fifteen or twenty years ago. Originally it was simply an outside thermometer. It's been taken to pieces. That's because nowadays there are small weather centres with electronic sensors that allow you to forecast the weather. So, this material had become outdated, that's why it's been replaced... It must be recycled because the liquid in the tubes is mercury. It's a toxic metal. The principle is based on a very old system that goes back to the first mercury barometers, maybe from 1600 or 1700. There are two measurements, one in Celsius and one in Fahrenheit. The temperature is always taken in the shade. The weather forecast has these international conventions. But I don't use it any more.

Thermometer, wood

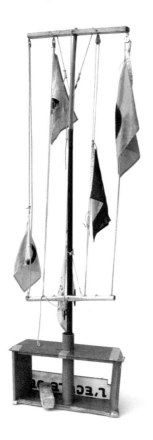

Creator unknown   Bordeaux, France. 1989

As recounted by Nicolas Dickets: We use this to train the kids, it's a teaching tool. It's something that they use with deaf people, because each flag has a meaning – so with these visual signals they know what they have to do. Each flag is different, and they each carry different meanings. The flags over there are international communication signs. For example, here's the letter 'T'. We can talk to each other using it, send each other messages. Then, when the wind blows a little, the flags can be seen. So the wind does the rest! We call it a signalling mast. Most boats have this kind of equipment. I use it to give my lessons, but it can also be used to give the signal to start a boat race, for example, or anything like that. Every now and then you have to change a string. It was here way before I was.

Wood, metal pole, pulleys, string, flags

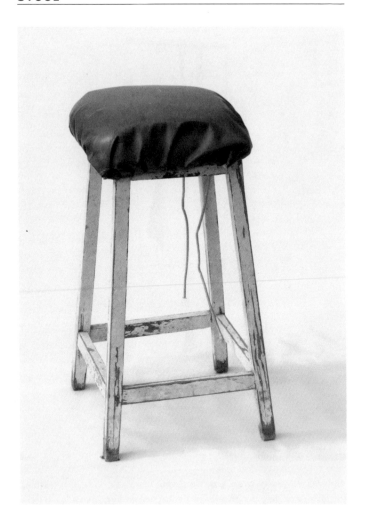

Creator unknown   Bordeaux, France. 1979

As recounted by Nicolas Dickets: We have a large boat over there, which was replaced. This stool originally came from the old boat. I've been here for seven years now, and this has always been around. This is foam. It's simply for sitting on. It's just a stool, a home-made stool. I use it for work every now and then, so I can sit comfortably.

Stool, foam, waterproof canvas, string

André Fernandez   Riberac, France. 2005

As recounted by his son Jérémy Fernandez, pictured: My father made it three years ago. He doesn't make things professionally, just for pleasure – his pleasure I mean! He made it for us, but we don't use it because it's too fragile, you can't sit down on it. We used it at first, but then we saw that it was about to break… He used something that already existed as a starting point. It's a chair for a small room, but really my father has plenty of space. He's got a lot of things he's made himself.

Wood, screws

Sarah Porfidio   Bordeaux, France. 2007

I made this about eight months ago. It's constructed from salvaged material. The wood was already joined together, I made these holes in a plastic plate with a drill, turned it upside-down and pegged these clothes pegs to the string. The wood was from a very beautiful wine crate. I made it because we refuse to buy too much stuff, and also it was just there, within reach. It's for the birds in winter. Birds come here, we've had couples here… Tits, robins… And two fat pigeons, unfortunately. When they come, we lower the roof, to put them off. We don't feed pigeons. There are some birds that like us, but we don't like them.

Plastic plate, wine crate, clothes pegs, string

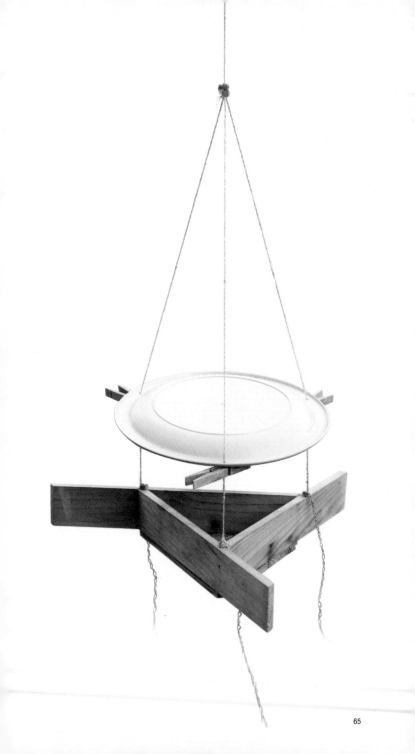

Renaud Beaurepaire   Bordeaux, France. 2009

It was useful because we drink a lot of beer and wine, so we used to make a lot of round trips to the recycling point. So I had this idea to attach a trolley to my bike, to avoid having to carry the empties. I found everything around here: my friend had used the wheels to move his boat, but because his boat was broken, he didn't need them any more. The trolley was already here when I arrived. It worked well, but I needed to put these wheels on so I could use it with my bike. I also took a metal bar and cut it a little bit. I made it over a year ago, we call it 'The beer trolley'.

Shopping trolley, wheels, metal bar

Renaud Beaurepaire   Bordeaux, France. 2009

These two parts were separate. This one used to be a stool but it didn't have any foam left, so your buttocks would hurt when you sat on it. We did everything up, tidied the place – it must have been just over a year ago now – everything was moved, and we didn't have any more space. So we put one piece on top of the other, and we put our computer there. This hole here is for the cables. There's no name for it. What would I call it? A computer rest maybe? It also had wheels on at first. We'd found them in the street and put them on.

Cable roll, round stool

Jacques Le Bars   Saint Gervais, France. 1987

These are irons from the early 20th century. They used to heat them up on the stove and iron clothes with them. I turned them into firedogs maybe twenty years ago. This one was an iron from 1900 that someone had given to us. I welded it there. They're for burning wood, so you can allow air under the logs and branches, especially the branches! And to fan the flames. It activates the fire. Because if you put your wood directly on the ground, there's no draw.

Irons, metal

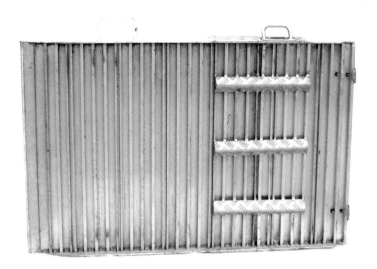

Jacques Le Bars   Saint Gervais, France. 2007

This is a ramp to lift heavy stuff onto the trailer. In fact this one was too short, so I extended and reinforced it. It's used to load and unload the trailer. When you want to put heavy things in the trailer, instead of carrying them, you can roll them around. It's a loading and unloading deck. An access ramp for the trailer and the wheelbarrow. Originally it was used to help disabled people in wheelchairs get onto the train, so it was pretty much doing the same thing. It's made out of aluminium, welded together. I had some spare pieces that I didn't need any more. I made it at work because we've got welding equipment there. Here, I could only use the blowtorch, which is more complicated. You can do it with the blowtorch if you use thin sticks of aluminium. I won't show you everything I've made because you might want to take it all away with you!

Aluminium

Gasser Firmin   Alsace, France. 1998

This is a dustpan, to shovel dirt. I made it while I was at a company called Loew
Keramik. It was fantastic there, I had a lot of time to work on such things.
Those were good times… It's a wonderful object, made from a very stable
metal sheet. It cleans the place rather well. I had spare metal that I welded,
bent and cut together. It must have been in 1998. I'm not sure what my
profession is, I do various things.

Steel

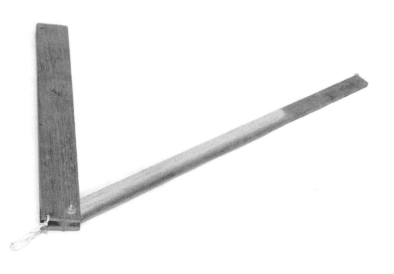

Emile Gbedemah   Bordeaux, France. 2005

The important thing is the size, the dimensions of the tool. This one here is for making cuts in wood when you're making furniture. But if you need to make a longer cut, you take a bigger tool. They didn't have one for sale anywhere, you see. I made it in 2005, and use it all the time. When you want to make a piece of furniture, you are given a place, a corner. So you have to measure. And this thing allows you to change the angle to make it sit straight to the wall. You put it here, adjust it, and draw a line. The easier it is, the harder it is to understand. We call it a 'bastard bracket' or 'grasshopper', like the insect. It's a bastard one because you can adjust it like that, depending on the angle of the cut you need to make.

Wood, nut and bolt

Fred Strobel   Fellbach, Germany. 1977

As recounted by Sabine and Giesella Strobel: We just call it the pineapple slicer, but it's a little miracle device. We're going to demonstrate it to you now! You can use this to adjust the size. This side is sharp so you can cut the peel off with it. This is the inner part, all the pineapple slices are instantly cut in the middle. My husband liked to eat pineapples, but back in those days – that's twenty-five years ago – unlike today, there were no useful implements to cut pineapples, and so our dad made this. Actually, he had nothing to do with metal in his job, he's an electrical engineer, but he wasn't afraid to use different materials. In those days all our acquaintances admired our pineapple slicer.

Steel, wing nut and bolt

Martin Brien Fellbach, Germany. 2003

It's a big adventure. I call it the 'self-made projector'. It all started when I was given an unwanted video monitor from school. I started to think about what I could make with it, and began work in 2002. The construction didn't take very long, but it was difficult to gather all the information to figure out what I could make with it, and how. My aim wasn't really to get a large picture (although, of course, I did want a projector, somehow – everybody wants a projector). But my real aim was to make something for myself. I used it for a year to watch films. The resolution isn't great, but you can definitely live with that. I watched the entire world cup on it. It's actually like a slide-projector, but larger. It's simply better! That's how it feels. You can buy a projector, but not in this form. The whole thing cost me eighty euros. In a shop, this equipment with this size projection would cost around 600 euros, plus the lamp. If I made it again now it would be much better!

Hardboard, plywood, video monitor, lamp

Gerhard Schmauder   Baden-Baden, Germany. 1978

I'm a botanist and a vegetarian, and I have a large garden, but watering it costs a lot of money. About ten years ago I cut off the top of a barrel to make a water butt, which was soon full to the brim, so I got a second one to use as an overflow. But I wondered how to get the water from one barrel to the other? That's when I had the idea. I thought I'd use an old vacuum cleaner tube. And so I made a 'water-levelling instrument', so to speak. I can't help it if it turned out to be a work of art.

Plastic barrels, vacuum cleaner tube

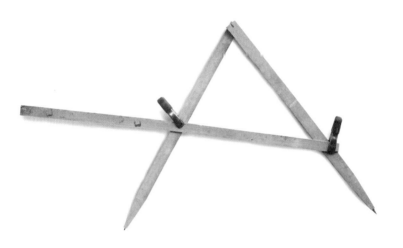

Bernhard Ruh   Baden-Baden, Germany. 2003

This is a pair of compasses I made to build the company signboard. You can use them to make geometric ellipses. When I designed and built this elliptical sign there were no compasses around with a dimension this large. So I had to build one, handcrafted like in the old days. I did it very fast: it took me fifteen minutes at most. Using this it was possible to construct an oval, elliptical plate. I did it in my workshop, and it's a very simple construction: a nail to join it, and two clip fasteners to change the radius. With this you can get a circle well over a metre wide.

Wood, nail, clips

Mathis Flotho   Karlsruhe, Germany. 2002

This is a Ferrari. I made it a year ago, but Dad helped me a bit. I painted it
with a red felt-tip pen. Dad gave me the wheels, this piece of wood I found
myself, and then I stuck on the exhaust pipe.

Wood, screws

Reiner Flotho   Karlsruhe, Germany. 2001

I found instructions on how to make this in an old book. I wanted to build it for my children, because I work with trains in my job – I'm a railway engineer. Although there is a strong tradition of model trains in Germany, I had to make this myself because you can't buy them for children to play with. There are only expensive models, but they're not meant for children, they're collector's items for adults. And this is fun… It's made from wood and painted.

Wood, screws

Reiner Flotho Karlsruhe, Germany. 2002

This is an imitation electric guitar for children. It has an amplifier too. This part is made from a bicycle spoke with a peg on top, the strap is made from a pair of braces. It's something you can't buy. My wife and I play flute and melodika. But not the guitar. That is still to come. The children are always making music, hitting kitchen pans with a stick, and playing 'air-guitar'. So I thought: let's make them a 'real' one.

Wood, bicycle spoke, braces

Reiner Flotho   Karlsruhe, Germany. 1994

I made this for my wife, when we were still very much in love but not yet married. It's a biscuit tin that she gave to me on my birthday. Inside there is an illuminated theatre. This is me, and that's my wife. It was a surprise gift. Back then she was acting in an amateur theatre group. When I met her, she was spending a lot of time at rehearsals, and so I had this idea. It's all hand painted and cut out.

Biscuit tin, paper, fabric, photographs

Eckart Pitz   Heidelberg, Germany. 1968

Before my wife and I were married we went camping in Spain. I sewed it all myself, it only weighs a pound or two when it's rolled up. I got the umbrella ribs in a repair shop in Erlangen, got them very cheap, and I cut them to the right lengths. This is aluminium conduit from a construction kit, I've bent and cut that as well.

Aluminium tubes, umbrella ribs, waterproof fabric, string, clips

Eckart Pitz   Heidelberg, Germany. 2001

First you need to vacuum the whole flat, then it'll be completely clean after you use it. I developed it years ago, but I couldn't get the angle right: it was either too short or too long. Then I got some plasticine from a handicraft shop – the sort children make small figurines from. I stuck it on and made it the right height so the vacuum cleaner tube can be pushed in.

Hair clippers, plastic pipe, plasticine

EGG WARMER

Klaus Jähnisch Fellbach, Germany. 1998

Actually, it's a hot-water bottle like the ones they used to warm beds before central heating existed. My grandmother used it as a hot-water bottle, and now we're using it as an egg warmer. You put the hot water in, and then you put the eggs on top. It's certainly very useful. When there are a few people at the table the eggs always get cold. And there are always several people at our table; on birthdays, or holidays when we invite my friends from the bowling club or when my children and grandchildren are here – then there are more than ten of us. We use this egg warmer two or three times a year. Of course you could buy an electrical egg warmer, but everybody has got one of those, and the cable gets in the way when it's standing on the table. When we put our egg-warmer on the table it looks fun, it's simple and doesn't need any electricity.

Hot-water bottle

FRUIT PROCESSOR

Hans Neef   Weiblingen, Germany. 1990

It's like this: mirabelle plums, damsons or cherries need to be chopped up to make schnapps. I use the drill with this tool that I made a few years ago. The tool is made of steel and there are small blades inside. I fill a small bucket with fruit, and everything gets chopped up with this tool. It's very solid, strong and you can't buy it in a shop. I've never seen one anywhere else.

Steel, blades, screws

Ralf Schmauder   Baden-Baden, Germany. 1999

It didn't work. I needed the tube to go up inside the car, to connect to the access hole. I had to customise it to make them fit together, because of the different sizes. I adapted the original lid, made a hole in it… It's just coincidence that the lid and the tube were the same size.

Plastic container, hose, clip

Viktor Agafonov   Baden-Baden, Germany. 2001

My father fought near Kursk. He was in tanks, and he was wounded. I used
to nag him a lot with my, 'Tell me all about it! Tell me all about it!' But he used
to keep quiet. I was a young lad, so I was interested in tanks, especially the
German ones – Tigers, and the rest of them. He never liked to remember these
things, but he used to watch me draw. Then one day my son got ill and I stayed
at home with him. He was bored, so I had to entertain him somehow. I
brought him some cardboard boxes and we started cutting the shapes out and
gluing them together, and we ended up with this tank. I was even more
interested in it than he was. He doesn't know anything about Tigers – he's a
different generation.

Cardboard, tape

LITTLE RED RIDING HOOD DOLL

Karin Rantzau   Ludwigsburg, Germany. 1980

This is a fairy-tale doll – it's Little Red Riding Hood and the wolf. Little Red Riding Hood goes to see the grandmother. The wolf eats the grandmother. And it ends up with the wolf lying in bed, when he's finished eating the grandmother. I made this doll twenty-five years ago, when I still had the patience. I call it the 'reversible doll', because you can turn it over. I had a friend who had one and I liked the idea so much! I have two sons, who were already too old for something like this, but I liked it so much, I just wanted to make one. It was a challenge! It has two heads, three faces and one body. I wanted to work it out! I made everything myself. But only once! Never again! It was madness! This is knitted, and inside the doll is filled with lambswool. First you have to make the body, then the heads, then the arms. All the pieces are filled with lambswool and sewn onto the body. After that the wool is used to make the hair, and then the doll has to be dressed. The eyes are painted on. First, stitches are made to create the recesses, and then painted. The forehead and the head need to be formed and knotted first. But my grandchildren are terribly scared of the doll. It's OK like this, but when I pull up the skirt and turn around the dress, another doll appears, and it scares them. That's why it mostly sits on the shelf now. It is not really meant to play with, rather to look at. It's actually more something for the grandmother, to tell fairy tales, like a puppet.

Fabric, wool, cotton thread, lambswool

Karin Rantzau   Ludwigsburg, Germany. 2006

This is a 'gateaux bag'. It's very practical. It's a carrier bag for gateaux and cakes, I always need to carry them. I can fit a large cake into it, and carry a large bowl on top as well, possibly two. You need to be able to put something on top. We are always donating cakes to the school or the nursery, or to the parish fête. I saw one somewhere and re-made it myself from pretty fabric, it's very practical.

Fabric

Rose Hajdu Zuffenhause, Germany. 1996

These are hobby horses. It's possible to buy ones like these, but I made these myself for the children, so that they'd have my 'signature'. This was the first horse for the eldest daughter and then, when the younger one grew up, I made the other one. There is an age gap of two years between the girls and they were three, four, five years old when they played with them. One of the horses is much lighter than the other. This is lime tree wood. This is lambswool. Wood is my passion, my hobby. I made many other things for children: animals, toys, many things. I still have some of the ones I particularly like. I put my heart into these horses.

Wood, lambswool

Horst Bauerle   Fellbach, Germany. 1987

As recounted by Ralf Bauerle: These are rear lights that are attached to this broomstick. My father made it at least thirty years ago. We just call them 'rear lights'. If you have no lights, this construction can be attached to the back of the trailer. The cable is plugged in at the front, and so you have lights at the back – if you're transporting a long ladder, for example. It's possible to buy the lamps, but not with the stick the way it is here. Nowadays you can get rear lights that attach using magnets and metal. Very modern. But there's no need to throw away something like this if it works.

Broomstick, lights, wire, reflectors

COBWEB BRUSH

Horst Bauerle   Fellbach, Germany. 1987

As recounted by Ralf Bauerle: This was made by my father to remove the cobwebs on the ceiling. Look, there are cobwebs up there somewhere. Yes, there! This makes it possible to clean them away quickly. It's very simple, and it works perfectly. My father made it about ten years ago. One could call it a 'cobweb remover', that's what we use it for.

Stick, rope, twigs

Ralf Bauerle   Fellbach, Germany. 2003

This instrument is for soil samples. The technical term is 'soil sample probe' but we didn't think about that, we just named it according to how it works – a soil extractor. You push it into the soil, turn it, and then the soil can be scraped into the bucket with a screw driver. This is a quick way to obtain the soil samples that we have to get examined every five years. Ten soil samples are needed for every vine area; then these are blended to obtain an average – because the vineyard soil differs from top to bottom. I take twenty samples from each of the individual sections of the vineyard. It's very important to take these samples, so I can find out what the vines need, and how I need to fertilise. There also are specific rules to ensure that the correct amount of fertiliser is used. It's like drinking wine: too much is bad, but too little is also bad. You can buy a similar instrument in the shops, but four years ago I simply took a water pipe, cut into it with a saw and welded it together.

Metal water pipe, welding

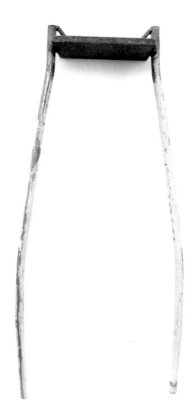

Ralf Bauerle   Fellbach, Germany. 2002

Back in the old days we used wooden poles to support the wires in vineyards. But the wood rots in the ground, and so these days we repair it with metal. I walk through the vineyards in spring to check where there are broken wooden poles to be repaired. The pile driver is needed to fix the poles into the ground. When the ground is too hard, then it's much easier to use metal. You need to use a lot of weight. First you have to pull it up to the height of the pole – they measure two meters seventy – balance it at that height, and then strike it down. You can buy pile drivers made of wood, but I made these from metal myself, just using water pipes. The thing is heavy, it must weigh twenty kilos. I made it four, five years ago. I can weld it myself, so it's cheaper.

Metal water pipe, metal plate

Klaus Burger   Baden-Baden, Germany. 2004

Pictured above. Well, I invented this for myself. I know there's some trombone player, who plays the Puccini and Verdi operas, he has a tuba here instead, but because this one has the sound of a trombone, it fits better. I had to play a piece of contemporary music by Helmut Lachenmann, and the musicians who had played this piece – a tuba concert – before me, had to play two instruments: one without a mouthpiece for 'air-sounds', and one with a mouthpiece. So I developed this. I've had this part from a trumpet built in here, and now I can vary the way I play.

Hose, trumpet part, tubas, plastic chairs

Michail Rojkov   Ludwigsburg, Germany. 2001

I made this when I was small, twelve years old maybe. I don't remember the exact year, it was when we lived in another part of town. I used it twice, on New Year's Eve when I made it, and again on New Year's Eve the year after. It started as a simple wooden construction, then I put this thing inside – a metal tennis ball container, so when you build a fire inside, the wood doesn't burn. And this is where I can hold it, so I don't burn my hands – because there are a lot of sparks. There are two handles, and this way you can launch and direct a rocket. Unfortunately rockets don't always fly straight, often they go crooked, so you have to be careful. I made almost all of it myself, sawed the wood myself, but Dad helped a bit. I call it a rocket launcher, in English. It sounds a bit cooler in English; in German it's called 'Raketenwerfer'. I later found out that it's quite difficult to shoot with this thing; it's less hassle simply to stick the rocket into a bottle and light it up. Well, I'd worked on it for so long that it would be a shame just to throw it away. It's generally forbidden to sell something like this, so you have to make it yourself.

Wood, tennis ball container, tin, screws

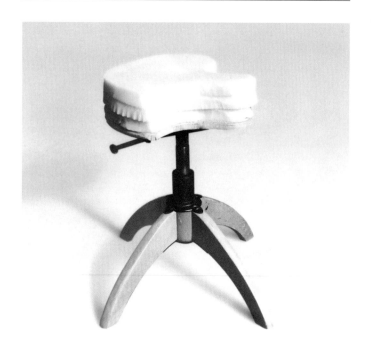

Matthias Winzen   Baden-Baden, Germany. 2004

It's very comfortable. It's good for the neck and it's healthy. I use it at the desk when I work there. It makes you sit in a certain way, which is good for me. I made it a week ago, to help with my posture, because I suffer from backache. They're too expensive in the shops, I don't want to spend the money. It's better if I can make it, then I can see if it works, and I can experiment with it. I can put more on here, make it higher, higher still… I made it using a knife, I just cut it, very simply. The legs are old, and the seat is new. If I bought it, I could spend all that money and it might not work. It's better when I find and experiment and then: aha! It's good. I had to fix my bed too, but I bought that in a shop.

Foam, stool base

Barbara Rau   Baden-Baden, Germany. 1990

As a student, I took a lot of photographs, but I never had enough money. The foam was taken from an old armchair, I can't remember where I got the material from. It's been sewn on the sewing machine, which is standing over there, that's an old thing too. It was 1992, I was studying at that time. On the one hand I had no money for such a big bag, but then there isn't enough space in a normal bag, you can't fit as much in. There is also the advantage that it is half round, which means you can carry it very easily. Normal photography bags are straight, made from aluminium, so they don't have as much space to put things in.

Leather, foam, cord

Tim Queally   Tulla, Ireland. 2004

I sometimes use buckets filled with concrete, they do the same thing as this.
I use them to anchor small boats when the wind gets up. I have bundles of
anchors, there are many different types.

Metal

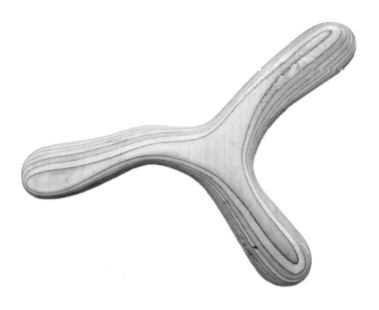

Aidan Moroney   Kilkishen, Ireland. 2006

As recounted by Pat McInerney: This is a boomerang that my brother-in-law made. He's actually a woodwork teacher, so he works with wood, he understands it. He made this about four weeks ago for my son, Austin. He had a plastic one, that he got in a shop, but he accidently broke it. He used it with Austin, they were throwing it, and it was coming back. The shape is unusual – they're usually like an L, but this works fine, it comes back.

Wood

Keith Towler Tulla, Ireland. 1993

As recounted by Pat McInerney: The clock was made twelve years ago, out of plywood, copper piping and various odds and ends. He made it as a challenge to see if it would work and keep time, and after various adjustments it did.

Wood, copper pipe, metal parts

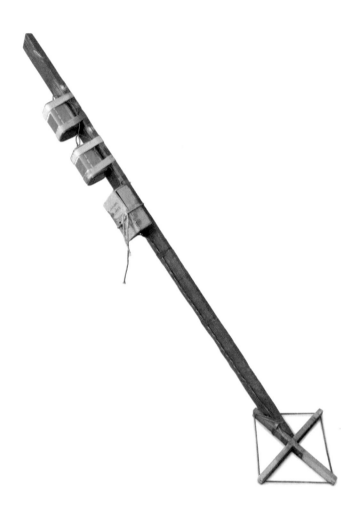

Keith Towler   Tulla, Ireland. 1994

As recounted by Pat McInerney: Keith made this because his brother-in-law
thought there were guns buried somewhere on his farm. He made this metal
detector from the mechanical parts of two radios, and it ran on torch batteries.
He fixed the device onto a wooden handle and set about trying to find them.

Radio parts, wood, batteries, wire

Michael Ryan   O'Callaghan's Mills, Ireland. 1988

As recounted by Angella McInerney, pictured: Michael Ryan was a first cousin of mine, he uses this for welding. See the glass is dark to protect your eyes. He made it from an empty plastic oil drum, it's just a bit of plastic.

Plastic oil drum, metal, screws, glass

John McNamara   Kilkishen, Ireland. 2006

You put it out into the lake. When you take a water supply from a lake you get little tiny shellfish and things getting into it, they clog up the water pipes, so this thing acts as a strainer. It's only a filter. Water has to soak in through that, slowly. The water is for farm use, cattle drinking. There's no design in it. It's simple, that's all it is. A big filter.

Canvas, string

John McNamara   Kilkishen, Ireland. 2004

Come on Bob, come back here! You're meant to stay inside! He lies in there and he puts his chin on the edge and looks out. It's a couple of years old. It was used for carrying oil. If I need something I look around and make it if I can. I forget what I used before, I think I had a barrel, a steel barrel that was chopped up a bit. If you buy then you have to put your hand in your pocket – that's the only reason! I'd like it if it looked better, but I think it's comfortable for him. It's not too hot or too cold I'd say.

Plastic oil drum

John McNamara   Kilkishen, Ireland. 2000

I use it at the top of a well. It's just a device to help me put pipes into a deep bore well that can be hundreds of feet deep. There isn't a name for it, unless you think of a name. I mean, when you make something yourself, you don't need a name.

Metal, car wheel

Phil Brown   Tulla, Ireland. 2005

I organise football every Monday and in the winter we play inside. But we have to take the goal apart every week. It's made from ordinary PVC plumbing pipes and joiners. It is nearly eight metres long. We only play seven or eight people in each team. For seven-a-side this is about right.

PVC plumbing pipes and joints

Gerard McNamara   O'Callaghan's Mills, Ireland. 2000

It was a quick fix that I did one time, I put a Christmas tree into it. I had a porter barrel, and so I put the tree on top of it. I wanted a bit of elevation for my Christmas tree once upon a time. I cut these notches and pushed this down into them. I made it in about five minutes on Christmas Eve! The old barrel was just lying around the yard and I gathered him up. It was about five years ago I'd say. It went missing for two and then it turned up again.

Barrel, metal

FEEDING TROUGH

Gerard McNamara   O'Callaghan's Mills, Ireland. 1994

The barrels were brought around by travellers, and we paid a small amount of money for them. I just cut them down the middle and made a frame for them. About ten or twelve years ago probably. It's a feeding trough, it's used for feeding the calves.

Plastic barrels, metal

John Culloo   Tulla, Ireland. 1996

It is a manual dispenser for barbed wire. You use it when you can't get to the
area you want to fence with machinery. Like a river bank for instance. My farm
has trees growing along the banks, you can't take a tractor in there, so you have
to unwind the wire by hand. This device makes that job safer. Your belly goes
in here and the strap goes around your neck.

Metal, plastic tube, leather strap, clips

Martin Rochford   Tulla, Ireland. 2006

Tis for the egg race, y'know. Tis only for messin', for racing eggs down a hill like, kind of a toy, just somethin' made up for the craic, y'know, for the young lady. It's only a one of a kind of a thing. There was a few more bits in it. There was an egg here, and an egg there. The heavier it is, the faster and the further it goes. They boil the eggs, put 'em in and – zzzooomm, down the hill! Tis only the odd time the race is held. There's every kind of a thing in it – shoe-boxes with wheels and stuff. I just kind of, I don't know, I just done it that way. I made it so it wouldn't collapse, fall over. And I had it kind of up high, so it would be easy to push it. Tis mostly all kids that enter, but I'm kind of like a big kid you know! Over in Feakle they had a soap-box race, I entered into that. I crashed the second time around. I sort of crashed for fun more than anything, I had my motorbike gear on, so it didn't matter.

Metal, wheels, bungee cord, feather

Maria Finucane   Tulla, Ireland. 1996

It was made ten years ago for Pat, so he could do a meditation course. Because he has a bad knee, he can't sit cross-legged, not for long periods of time. With this you can sit with your legs under here. See? You can sit for a long length of time and meditate.

Wood, screws

Maria Finucane   Tulla, Ireland. 2003

You can use them over and over again. If I was going back into teaching, to
teach printmaking… I've these kept for that purpose but… I would store them
like this in the box of tools, with my ink, with whatever, so that when I put my
hand in I don't get – !

Wood, tape, metal, polystyrene

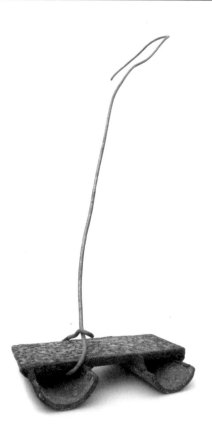

Bernie Whelan   Maghera, Ireland. 1996

It's made to extract nails from sheets of corrugated iron, without damaging the sheets – so you can use them again on a new shed. It fits into the grooves of the iron, and you can push the crowbar against the metal without it touching the sheet. It's a bit crude but...

Metal, wire

Ray Egan   Tulla, Ireland. 2005

I suppose we made it about a year ago. When we were making this place here. This is an old plastic barrel that would be used for holding chemicals or oil. They're very common, you see them everywhere. Normally we use them for water. This is for cement, for concrete. The reason we made it is because we have a small machine with the forklift, and it's easier than using a wheelbarrow. I got the idea from the building sites, they have big metal ones. I call it the half-barrel for carrying cement! Because it's plastic, it keeps the water in the cement, so it keeps it fresh... Well you see we have a mechanical mixer that has wheels, but it's too hard to move around.

Plastic barrel, wood

James O'Hagan   O'Callaghan's Mills, Ireland. 2001

It's just a barrel stand, the holder for the barrel. The oil companies who sell barrels of oil would sell you that type of thing with the barrel. I think they were giving them away free at one time. But theirs are a little low, so I made a higher one. A standard one, you can only put a five litre container underneath it. I can put a twenty-three litre drum under that when I have a tap on the barrel. When the barrel is full it's hard to get it on. I made it myself about five years ago, from material in my own workshop. The shape enables you to lower on the full barrel without difficulty. You see, it can roll on. It has a rolling effect. But somebody else invented the original invention. I just made one for myself that was higher.

Metal

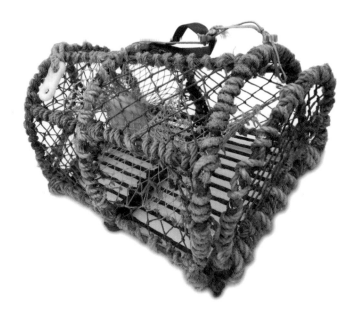

Digger   Liscannor, Ireland. 2004

Lobsters are scarce now. It's still early in the year, but there's too many
fishin' them now. It should be a licence job, and give it to the fishermen – not
farmers. There's more farmers fishin' for lobsters than there are ordinary
fishermen. The concrete is used to hold it on the bottom. The frame comes
from Scotland. We do the rest ourselves, put the net on… That's my boat
out there.

Metal, rope, net

Pat Minogue   O'Callaghan's Mills, Ireland. 1982

It's a rocking motorbike. I decided not to make just a motorbike but a rocking motorbike. It's twenty-four years old, and it works! I got the basic idea from other toys like this. But the motorbike... It was very popular with all the children who came to visit the house, and even though they might have more sophisticated toys: bicycles; machines; and the like; this was more popular for some reason. Because after a little while you got good at sitting on it, and then you got good at doing stunts on it, and then...

Painted wood

BOTTLE CORKER

Pat Minogue   O'Callaghan's Mills, Ireland. 1991

I used to make beer, home-made beer. I made this to fit the right size bottles.
I would just leave the new cork there and just go - BANG! It's for putting tops
on Belgian beer bottles. This is my own idea. I suppose you could call it a vice
for putting corks on a bottle? No, it's too long. A bottle corker? It's about fifteen
years old. I've since stopped making beer and started making wine.

Wood, screws

 **Lavender Wright**   Tulla, Ireland. 2001

All you would do is kind of hang it on like so. Then you would put your hands in. You have your clothes and your pegs. Like this. Then, sure you'd start hanging...

Fabric, hanger

TOOTHPASTE TUBE SQUEEZER

Fred Becker Tulla, Ireland. 2006

When toothpaste came in metal tubes you could squeeze out the very last bit. But when the tubes changed to plastic it was impossible to squeeze all of the paste out. So I invented this device.

Wood, nut and bolt

Fergus Tighe   Liscannor, Ireland. 2001

I made this postbox because the door of the house has no letterbox. It's a big job to make a hole, and in the wintertime the wind from the sea is like, 'WHOOOSSHH!' So I went to the store, to see if I could buy a postbox. They have them, but they cost sixty euros. I'm no good with a hammer and nails, but I made this. It's broken but, y'know, it's very good, You have to put a big stone inside, in case it blows away, in case the wind takes it away. It's a thing of great beauty. The paint was very cheap, as you can see I painted the door and there was some left over. I did a proper job, primed it. But even though I put primer on the paint still came off. I'm surprised. I made it just a year ago. I lived three miles away, and when I moved they kept my post at the post office. But the man in the post office got pissed off, he said, 'You have to get a postbox!' I said, 'Sixty euros for a postbox! No. I'm a poor artist!'

Wood, stone weight (not shown)

Kerry Hall   O'Callaghan's Mills, Ireland. 2006

I went to the shop and he had this old bucket. I said, 'Oh, that'd be handy for the post.' My friends living outside Tulla, they all use buckets as well. My mother came over and she wanted to buy us a new one. We said no, the bucket's doing the job, it keeps the rain out. Maybe when the house is built we might get a new one.

Plastic bucket

Jim Murphy   Kilkishen, Ireland. 1998

It was for bringing milk to the creamery at one stage. I was milking cows but I wasn't supplying it in that. I made most of it up. I got the holes punched out and put those teats around them, and when you put the milk in there, the calves went and sucked it. You could buy one – but what are you paying for when you're buying? I had most of it anyway. All I had to do was weld a few things and get that punched, you see? Most of the value was in it already.

Trailer, stainless steel container, metal sheet, rubber teats, hose

Jim Murphy   Kilkishen, Ireland. 1987

I call it a paddle. This thing over here is as near to it as makes no difference – it's a bit like this thing, but that's not exactly it. Because if you put that under the silage bale, it'll cut the plastic. Yeah, so, we put this over it. Like this. Now do ya see? See what you have now? The bale sits on it, and the plastic's not cut. It must have been made – oh Jesus – about fifteen years ago! Whenever the bales came out.

Metal

Brendan Finucane   Kerry, Ireland. 2004

It's a probe used to help measure the depth and angle of boreholes for explosives in stone quarries. This part is a real baby's bottle! We designed it about two years ago. It's about the fifth one I've gone through. You put a torch into the baby's bottle and you drop that down the hole, so you can see the light. When that's down the hole, I use my measuring instrument to check the angle of the holes in the blast. Sometimes the actual slope of the rock necessitates that you have to drill at an angle. If it's drilled at an angle, you drop this into it, and it helps to tell us what angle it was drilled at. Let's say the driller is told to make all the holes at ten degrees. Then we use the torch. We just throw it down to the bottom and we can see the torch there. Once we look through the measuring instrument, we actually know – this actually tells us – whether they are at ten degrees or not. There's a probe you can buy, it's what they call an in-hole probe. But sometimes the rock is loose, and when you throw that down a hole a rock will go down after it, and you can't get it.

Tape measure, baby bottle, torch, tape

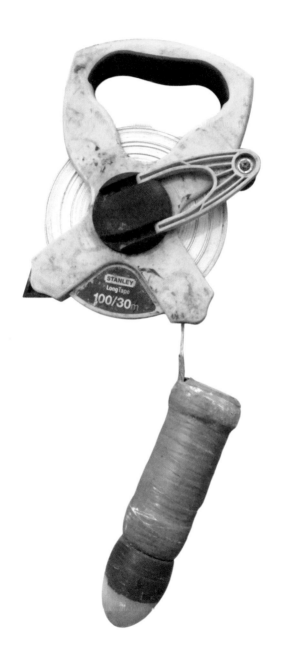

John Joe McNamara Bodyke, Ireland. 1995

It's a turf barrow. Specially designed for turf. It took about fourteen hours from start to finish, roughly. There's a lot of hard work in it… It was made specially for this job, you wouldn't do anything else with it. This wasn't my first time making a barrow. I had it in my head, the plan of it. When you have it all in your head, it's like a man going and making a horse box. It's all in his head and he knows what to do.

Wood, screws, nails

Derek Wright   Tulla, Ireland. 2005

A lot of the bird box designs, aren't really right to me. A lot of them have the stick coming out under the hole. They shouldn't have that, because that means that another bird can perch there and look in.

Wood, glue, nails

Victor Tsvetkov   Quin, Ireland. 2001

The important thing was that you could always take it apart, then put it together again, you know? I go shopping, they want four hundred pounds for these. I says, 'For what?' I have hands. I can spend the money on tubes of paint.

Wood, nut and bolts, castors

COMMUNAL ASHTRAY AND FILTER

Giancarlo Cencetti   Prato, Italy. 2002

This object was initially created to put cigarettes out, we put sand in it and used it as an ashtray. The holes were punched through the metal 'cup', so you can take the cigarettes out of the container, and you don't have to throw them out one at a time. It's a 'filter'. It was created about a year ago, so that cigarettes are not thrown on the ground. We don't use a 'normal' ashtray because they go out better in sand. It's safer. I had the idea to make it, it hasn't got a name, I don't know if it exists anywhere else. We just used the material that we had.

Oil drum, sand, tin

Odino Tumiatti   Milan, Italy. 2001

It's a machine for dividing rolls. I used to buy large rolls of paper at wholesalers or in supermarkets where they sell those kinds of things. It's for making them smaller, for home use, for cleaning your hands and so on. You see, this one is larger, and I want to make smaller paper rolls that are this size. I put this roll inside, on top, and one underneath, so one winds around while the other one unwinds. See, here I've got the lathe, I hang the roll up here and it's practical when I have to clean my hands. No, it hasn't got a name. I made it some ten or fifteen years ago, but can't remember exactly.

Wood, metal rods, nut and bolts

Odino Tumiatti   Milan, Italy. 1983

I made this one here. It's for pounding dry bacalao – dried cod fish – not the wet ones. When you buy wet bacalao, it's been put in water for six or seven days and it gets ruined because the water takes away all the taste. When I buy bacalao, I sprinkle it with water and I pound it. It becomes crumbly and doesn't break. Then I cook it and it comes out really nice because it hasn't been washed. Yes, I made it myself, but I make all my stuff myself. This one must be fifteen or twenty years old.

Wood, plastic straps

Mario Pieraccini   Prato, Italy. 2000

I made it when the dogs were born – four years ago. It's their bed. It would
be bad for the dogs to sleep on the floor because of the humidity. I made two
of them – one for each dog, but since they're always together, they only use
one. It's made with iron and fabric. Recycled stuff. I didn't buy one because
I like making things myself.

Iron rods, nylon sheet

Mario Pieraccini   Prato, Italy. 2002

I made it two or three years ago. I put charcoal there and cook steaks, but now I don't use it any more. It's old: there are better ones, but this one is hand-made! There's one fixed in the wall, but can't be moved around. This one, I made for the warehouse, it's broken, all rickety. These bits on the top are used to put the grill up depending on the type of meat that you're cooking. beefsteaks need a strong fire, chicken less. It's very craftsman-like. The wheels are from an old vacuum cleaner.

Iron sheeting, iron rods, hose pipe, metal grill, plastic wheels, plastic basket

Alessio Cecchi   Prato, Italy. 2002

I glued a five cent coin to the mobile phone and fixed a magnet onto the car's dashboard for convenience. I put my mobile here, press a button and talk on the phone. It's very simple. I invented it myself. I think I've had the magnet since I was twenty years old. I'm never in the car for more than twenty minutes, and it's annoying to put my earphones in and then take them out, and in and out again. So this solved the problem.

Euro five cent coin, magnet (not shown)

Bruno Ballini   Prato, Italy. 2005

My name is Bruno. I made this container a year ago, to make cleaning easier. It was a barrel for liquids, originally. I made it with a cutter. It's much more practical when it's made like this because you can empty it better with one handle, and it's less tiring. Inventing it was easy, we just had the idea to do it. We only use it here. I didn't give it a name. I made it because it's useful.

Plastic barrel, plastic handle

Adelmo Pieraccioli   Casa, Italy. 2002

As recounted by his wife, pictured: This is a small bench. I have another one as well. I always use it in summer to sit outside. My husband Adelmo didn't buy them because he liked making things himself. I kept them in the attic, but then I thought I'd put them outside – they could come in handy.

Wood, nails, cloth, string

Mario Cifliku   Cartaia, Italy. 2007

This thing here is a goal. I made it with wood and nails, and then anchored it with breeze blocks. I made it two months ago. You can't play in the main square, so we moved here and I created this goal to play football.

Wood, nails, breeze blocks, bricks

Luca Targetti   Prato, Italy. 2003

This object is used to produce warm air. When the fire is lit, air goes in one pipe, is forced through other pipes, and then at the end warm air comes out. I used an engine from an old wall-mounted air-conditioning unit, a convector. It's also possible to use this to heat water. It didn't work well at first because the fireplace had bad suction. I made it with my father.

Iron piping, motor

Dante Santini   Podere Porcile, Italy. 2002

I was using this thing before they invented pumps to clean drains. I made it some four or five years ago. There were bigger ones, but this one is enough for me. It's called the 'Tozzino'. I created it with a cylinder and made holes in a zinc-covered saucepan, which doesn't get rusty. I didn't buy one because this costs less, and I like it better. It's more practical.

Saucepan, stick

Angiolo Boanini   Casa, Italy. 2006

This is a rat trap. Something ate through the wooden board once and escaped. I worked in textiles but I've always liked hunting, it's my hobby. I go rat hunting as well! You can find smaller traps in shops, but the bigger rats won't fit into them. The traps around here are all hand-made. I use this one here because the rats eat the pine nuts that fall from the tree.

Wood, metal rod, rubber inner tube, net, hinges

Sven Hünemörder   Altai region, Russia. 2003

My wife and I were on our honeymoon in our Siberian homeland, visiting my
parents-in-law. My wife was making a film and I was responsible for the
music, but I couldn't bring my guitar with me. I didn't think it would be so
beautiful and inspiring out there. I locked myself in the barn toilet, and carved
a guitar neck from a scrap piece of wood. My father-in-law knocked on the door
and shouted, 'Are you all right? Come out!' But I was searching for something
that I can use to fix the neck to. In the same barn, I found this drum, which
used to be part of an old Russian tape-recorder. I made the holes with a
screwdriver. By the time evening came, I had an idea for a musical composition.
At the premiere of the film, my wife was supposed to say a few words, but I went
up there instead and played a little something on this guitar.

    At the only music pub in Tashtagol (Province of Kemerovo, Russia),
Zhenia (my wife) bet me that I wouldn't be able to play *Hotel California* on
it. I won the bet. You tune it like a guitar, but it has four strings. So is it a
balalaika, or guitar? We call it a balalaika, but it's played differently.

Metal drum, wood, scrap metal, guitar strings

Alexander Sakovnin   Vereya, Moscow region, Russia. 1960s

As recounted by Galina, pictured: Our dad's friend made it, it's called the Little House. Dad passed away a long time ago. His friend was making it for his daughter. And when she grew up, he offered it to us, 'You've got two more growing up, there, take it!' And so we rolled it, this little house, on logs down the lane. We set it up here. So how long has it been standing here? I'm forty odd years old now, so this little house must be at least fifty. The man's name was Sasha – Alexander Sakovnin. He's not alive any more, but his Little House remains. You can see the pictures I stuck to the walls inside, covering over the original wallpaper. These other pictures are a bit more grown up, someone else glued them up. It's always been known as the Little House. When my children were growing up they always used to say, 'We're going to play in the Little House!', or 'Where are you? I'm in the Little House!'

Wood, sheet metal, glass

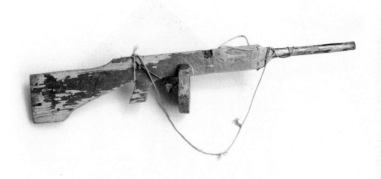

Vladimir   Kolomna, Russia. 2006

I made it for my son about three years ago, not for any particular reason, just to play with. Then he grew a bit older and they made him one with little bullets. If I was making one with bullets, I wouldn't make it like that – you have to make it with a straight stock. It's not a Kalashnikov, nor a PPS, it's a hybrid of some sort. He made one for himself later.

      He saw these guns in American action films. He showed them to me, I looked at it and knew it would be a piece of cake to make. There are plenty of toys like this in the shops these days, but he doesn't want any of them. He's been running around with this wooden thing for three years now, and if you bought him one from the shops he wouldn't want it. He's got plenty of toys, you name it, he's got it.

Wood, string

Kostya Kuznetsov Yaroslavl region, Russia. 2003

The rooms are small, and we had a baby, we had to keep him separate from the kitchen. I mean, the bedroom's really small. So we'd put this up, unfold it. We only needed it for six months, after that he'd start opening it himself – then it wasn't any use. We only used it until he was about eight months old. Then he started walking. Before we made it fold like this it would take up a lot of space when it was open. Like this, when it's folded, it's not in your way. I sawed it in two, and that's it. I'd seen something like it, a foreign one, but it was on casters. I wanted to buy one like that, but I couldn't find any. We use this in the kitchen and everywhere really, it's very handy. I don't know why they don't make ones like this in Russia. I just had to make one myself.

Wooden door, hinges, tarpaulin

Valery Stepanov   Moscow, Russia, 1990

As recounted by his son Andrei: Gorodki (a variation of skittles) is an old Russian game. It was played by Tolstoy, Lenin and Pavlov. I used to watch my father play and gradually started to like it myself. I was just watching, learning to recognise the sounds of hard and soft hits, listening to the hissing sound of the bat sweeping the ground and how the *gorodki* bounce off. When it would go 'Bang!' on the ground, that's the wrong hit. I got a feeling for when it was played well. I thought I'd try it myself, and in the early 1970s my father took me to the Dinamo stadium. The coach gave me my first bats. He showed me a large crate with forty or fifty pairs of bats in it, bats are always made in pairs. He chose me a pair to suit my height, and that's when I started to play, I was about ten. Gradually I reached semi-professional level and learned how to make the equipment. The bat is a complicated tool consisting of six 10 cm steel bindings. The last binding is called the 'glass' because it's the shape of a glass. The bottom of this glass is very thick – you need that to shift the centre of gravity. This gives an advantage when the bat is turning. The stick itself also consists of small pieces of 15 cm wood, 5 cm of which go into the steel bindings. So the whole stick is assembled and made of wooden fittings into steel sections. The handles are made out of dogwood. Practice has shown that there is no other wood that can endure as many hits as dogwood can. That's why *gorodki* players will travel all the way to the Caucasus to sneakily chop thick dogwood.

Wood, metal

Andrey Repin   Moscow, Russia. 1984

This is my favourite kettle, I bought it in Severodvinsk. I think it was made out of titanium from a submarine at one of the defence plants there. It still works. So many times we've forgotten to keep an eye on it. It doesn't switch off automatically of course, and when it overheats it goes red, it gets so hot. The spout got lost somewhere, it's legs fell off, and eventually the handle too. But it was working fine, it would have been a shame to throw it away. So I nailed a wooden piece here, with a hole drilled in it, for the water. The legs were plastic, they broke off. So I soldered pieces of metal on, but one of them has already gone. It's a fantastic kettle, we've probably boiled an ocean of water in it. I think the handle was taken from a rake, but I can't remember exactly. The wire on the front is just to secure it, because it's not properly fixed in any way, this makes it more solid. Of course I could buy one, but the way I see it a kettle is like a friend, a life companion. It'd fed me since 1984, how could I throw it away? You usually throw them away when they burn out, but this one never dies.

Titanium, wood, wire

Aleksandr Yuzhnoye Butovo, Moscow, Russia. 2001

As recounted by Boris Petrovich, pictured: There's not much to tell, it's a summer shower. Well, the man who lived next door worked as a bus driver. There were loads of old broken bus doors at the bus depot, doors that had been written off. So he decided to use them. He put them together here, fixed them together to make corners, like this – so that they stand up, they don't move. They're heavy, sturdy, they stay up. That's it... No problems here. It holds itself. Two crossbars here, and a barrel on top of them – the tank. When everything's heated up, just open the tap and shower...

There's no hot water here, soon they'll knock us all down! They keep turning the electricity off. They threatened to turn off the water supply too, so there's no point in doing anything properly here.

Bus doors, iron bars, iron barrel

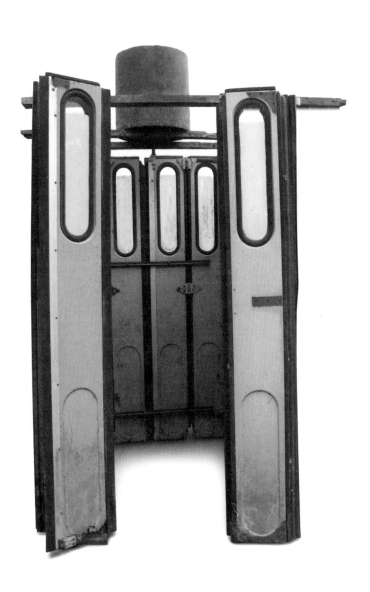

Alexander Kotov   Ryazan region, Russia. 1998

As recounted by his grandmother Nina Sergeyevna Kotova: When you make felt mittens, the wool becomes ugly and crinkled. Then you put them on the template and it straightens them out. But if you make gloves you have to do the same thing to the fingers. My grandson Sasha made these for me. 'Grandma,' he said, 'let me make you hands so you don't have to go through all this trouble.' I used to put the gloves onto small sticks, but they kept falling apart. So I said, 'OK, give it a try, it might work out.' To start with he simply cut the four fingers out of a wooden board, and for the thumb we used a stick. He probably didn't manage to make one with a thumb at first. Then I suppose he thought about it for a while, and cut out this shape with all five digits. Of course it's better like this – the glove fits like it does on a hand.

Wood, tape

Olga   Yaroslavl region, Russia. 1993

Full name and picture withheld at the creator's request. Well I was stupid, I
stole something and they locked me up. I was in prison in the Yaroslavl
region, sent down for five years. We used to sew slippers and mittens. I was
young so I was frisky. And where's a girl to go without a man? It was real torture,
you'd start humping any stick. A carrot or a cucumber – everyone does that.
And what happens when you want one and there aren't any around? So I found
a piece of rubber and used that. Fuck knows what kind of rubber it is, I just
carved it up with a knife when nobody was looking. I still keep it in case it comes
in handy again!

Rubber

Viktor Nikolayev   Loukhovitsy, Moscow region, Russia. 2005

It's a headlight. It's not so strange looking – it just turned out like that... My brother lives in the Rostov region, he does car repairs. For a long time I'd been asking him to sort me out a car of some kind. Not an expensive one, but in good condition, something I could touch up a bit and have myself a decent ride. So he called me one day and says he's got a good one, and to come and get it. I went there to have a look at it – he was right, not too much hassle, and the price was reasonable because the owner wanted to get rid of it. Some bits were missing, but it was a piece of cake. It had been in a crash, and the front was damaged, there was no headlight. I didn't have that far to drive, but it was dark, and then there are the police...

My brother was doing some work on his house, so we took these halogen lamps, and an empty tin. We cut out the base of the tin to fix the lamps onto, connected it all up, and drove back just fine. Then we took it out, fixed it all up and bought the right headlights.

Tin, halogen lamps

TUNING FORK

Nikolai Timofeyevitch Proskurin   Norilsk, Russia. 1994

There was a time when my second profession, alongside managing a symphony orchestra, was tuning pianos. I had a tendency to make everything myself, with my own two hands, especially during the perestroika period when there was nothing like this in the shops.

These are the tuning forks, which I made myself. This one is brass. I used to cut them out from blocks of metal using a hacksaw, then polish them up. Then I'd tune it properly to 440 Hz. They're different from manufactured ones, they would only sound for fifteen seconds after impact – mine would sound for a minute. So the tune of the starting note 'A' comes out much more precise. When I arrived here, it turned out they needed piano tuners too. I made a few more tuning forks, I already had the material because my wife is also a piano tuner. we both needed them. But now there are electronic tuning forks, which are more precise and practical to use, and they show any frequency.

Brass

Albina Leonidovna Falko   Perm, Russia. 1985

As recounted by her son, Mikhail Turbinsky, pictured: Our school was small – only two classes in one year. And some time in year 4, around 1984, they put in two table tennis tables. I was at the age where I liked to learn to play table tennis, and was old enough to chase the younger kids away. It was hard for us to get anything back then: the bat I had was the one my mother had played with in her time – back in the 1950s when they were young. It's a real Chinese bat – very light, but the rubber made in 1958 had turned into wood over time, so it was impossible to play with.

The first thing I did was to beg my mother for a piece of rubber (she nicked it from the factory, everyone was stealing stuff in those days). We sanded the bat with an emery board and stuck the rubber on top of it. Of course, the rubber didn't allow for any spin or other effects. But as our playing technique began to develop, we needed to start spinning the ball. The solution was perfectly simple: you buy a balloon and stretch it over the bat. This would give you enough stickiness on the bat for about a week. Sooner or later it would naturally tear, then we'd put another one on.

But there was an additional problem… The table tennis balls. What did we use to carry them in? Well, our school bags, naturally. It's obvious that if a heavy book falls on a ball it will dent it. The balls were cheap but practically impossible to find in shops. They were valuable things. We had to be careful with them. I remember that if someone accidentally stepped on a table tennis ball but didn't break it, there was a special method to rectify it. You take some boiling water, drop the ball in it, and the air pressure would spring it back out from the inside. The ball would take a slightly oval shape if you didn't perform this procedure carefully enough, and there would be some indentations left where the ball had been creased, but you could still play with it nevertheless. One day I complained to my mother saying that it would be nice to make some sort of a small box so that the ball wouldn't get damaged, and as a result I got this wonderful piece here. It's made from titanium. On top of that it was argon-welded together with some kind of aeroplane material. It's just a small box. The ball fits in there perfectly. But the lid was constantly coming off in my school bag, so I had to transport it with a black rubber band across it.

Titanium, rubber band

TOY STEERING WHEEL

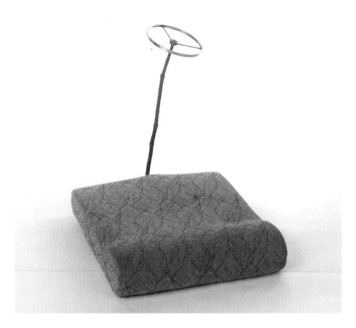

Albina Leonidovna Falko Perm, Russia. 1978

As recounted by her son Mikhail Turbinsky, pictured on page 158: When I started working at the Sverdlov factory in 1988, I used to turn this sort of thing myself. It's some kind of part from a TU-134 engine. This unwanted metal collar is used to make a steering wheel for a child. When I was a kid, we had a small sofa. I'd take one of the cushions off and sit on it and steer, like I was driving a car. In the 1980s there were very few toys that actually had steering wheels, and those that did were too expensive for parents to be able to afford them. So this was an excellent alternative for a child to be able to 'drive' at home. My son now plays with a real steering wheel from a *Kopeyka* (a VAZ-2101), but back then it was a useful piece. It did the job just fine. I was still in kindergarten when my mum made this for me, so it must have been in 1977. One day she brought home a load of different metal collars and ballbearings. They were probably taken from the bucket where they put the metal that's to be remelted. I have no idea how she managed to nick stuff from such a high-security plant, but I was very happy. There was this pile of interesting metal that I could play with, making all sorts of things. She got fed up with me nagging, begging her to get me a steering wheel, so she quickly put this together for me so I'd get off her back. It's a five minute job for an experienced welder.

Metal, wood, cushion

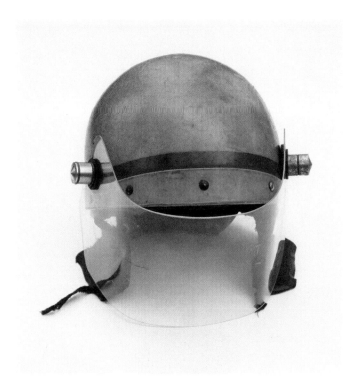

Petukhov Leonid Ivanovitch   Moscow region, Russia. 2003

I use this helmet when I drive my tractor, because my head gets cold. This model should have had a peak, but it broke off over time, so I would drive without a peak or a visor, but then all kinds of rubbish would get into my eyes. Also if you are driving in the cold weather, your face gets frozen – especially your nose. So to avoid that I made this visor, I bent the perspex myself. These are plungers from syringes, I use them to move the visor, I simply pull them out, lift up the visor and push them back in, then it stays in position.

Helmet, syringe plungers, perspex

Sergei   Tolyatti, Russia.1998

A present from the heart. It was all cut out and painted inside to give it some colour. See the red here it was touched up – green, blue, black. Yes, you can take off each ring of the stand separately. And then I fixed these sheets made of clear perspex, and glued the fabric on top.

Perspex, fabric, metal, wire

Nikolai Viktorovitch Karpov   Moscow, Russia. 2001

Well I didn't do much here. It's just a normal stool. To be honest, I grabbed this bit from the rubbish tip. I guess they were refurbishing the cinema, and they threw it away. So I picked it up. It's just a simple chair without a seat. But I didn't need a seat – you see? I have stools in the kitchen here. There's no space for anything bigger in this room. You want to have some comfort, but there's no space. Now I don't have to lean against a cold wall. You can't think about anything else in these conditions, it's as if it was made to be! I was walking past the tip and saw these nice red, soft armchairs. But they were all broken: you could either have the back without the seat, or just a seat, or even just the legs on their own. Maybe the workers broke everything when they were pulling them out, or maybe tramps. They were dirty, so I picked out the cleaner ones. That's it – nothing else to tell.

Cinema seat, wooden stool

Vassily Grigoryevitch Arkhipov   Kolomna, Russia. c.1965

There were no fridges at that time were there? We needed one for the family, and it was impossible to buy them. So I decided to make a fridge myself. The reason was simple: necessity, of course. I got this opportunity as I used to travel a lot for work. When I was in Moscow I saw that they were selling refrigerator motors in one of the shops. I had a close look at different fridges to see how they were put together. At that time the most famous make was the 'Saratov'. But they also had the 'ZIL', and the 'Oka', but we couldn't even dream of those back then: they were very expensive and rare. But they sold fridge parts, and the individual price for these parts was reasonable. So I decided to buy this part in Moscow. It was sold in this heavy wooden box, far too heavy to carry. So I took it all out and left all the wooden bits, because it was easier to manage the metal parts alone. I packed it up and brought it home.

At the time we had a builder staying with us, I can't remember his name. He offered me some plastic. I'm not sure what type of plastic it was, but it was easy to shape using the right chemicals. I took three sheets of this plastic – I don't know where he got it from, it was probably stolen, or perhaps there were some leftovers – and I glued them together. On the bottom sections and where the shelves are, I reinforced it using metal. I put extra sections in the corners, especially at the bottom because I stored nine three-litre glass jars there. I had nowhere else to put them. We made gherkins in autumn, and pickled tomatoes. I kept them in the fridge on two levels, as we didn't have a cellar then.

Wood, plastic, refrigeration motor, metal

Sergey Vassilyev   Stolptsy, Ryazan region, Russia. 2008

It was 2008, I was visiting my mother. It was only autumn, but it had already snowed, and I couldn't find a snow shovel. Well, I found a sieve from a winnowing machine and thought it would make a good lightweight shovel. I cut some bits off the sieve and the machine, and made this. It turned out light and very easy to use. It's a metal shovel, so it's quite solid.

Sieve, wood

Petr Naoumov   Staritsa, Tver region, Russia. 1994

I'm not going to talk about it, what's to tell anyway? It's obvious, isn't it? I made it for my son to play with. He likes it… I made it with whatever I had at hand. The trunk is from when we knocked the apple trees down, there were branches lying around. There's some plywood, the wheels are taken from his own pram, from when he was very little. I nicked the wire when we were pilfering from our collective farm. I like it myself… I dreamt of one of those when I was a child, so I made one for Kolya. They like lying on the ground behind it and shooting. They get dirty like devils after that though…

Tree trunk, pram wheels, wire, plywood

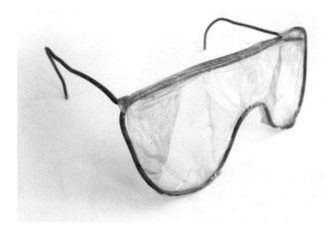

Boris Borisovitch Knyazev   Moscow, Russia. 2005

I needed to do a job really quickly. I was cutting steel with a jigsaw. Well, the dust would go straight into my eyes – but the job was urgent. It was a weekend and it was physically impossible to get a pair of safety glasses anywhere. So I just took a piece of wire, some tape, and made these goggles. Literally in no time. I had a cup of tea and back to work again. Metal dust gets all over the place, not like the wood dust. It would get me in the face, in my eyes, so I had to squint. I would be squinting, and the job was also quite delicate, well, more or less delicate... But it would still be annoying. So instead of squinting, it's better to keep your eyes open but through a layer of film. It's just thin cellophane. It was only a one-off job, sawing those corners, so it makes no sense for me to buy glasses for this. All I had to do here was to think of how to make them, there was nowhere I could have looked for them anyway. So I just thought of something, put it together and carried on working. This is steel wire, galvanised, I think. Simply rolled together, bent and fixed with tape. It was made in spring. Since then I'd put them down and they stayed there, waiting for their moment to shine.

Wire, tape, cellophane

Mikhail Akhmatov   Moscow, Russia. 2003

I bought this photograph glossing machine a long time ago. A good, solid, laboratory thing. They knew how to make certain things in this country back then. It would dry photographs in a single turn of the drum: they'd become glossy, smooth and beautiful. I haven't had a lab for a long time, technology has moved on.

A while ago I really needed to mix a large volume of cement properly. I remembered this was sitting in the garage, and thought I'd turn it into a cement mixer, so I didn't have to use a spade or a shovel as much. The whole construction is similar in principle anyway. Unfortunately, I'd thrown away a lot of its components – the electric engine, the gearing and so on. So now it can only work manually, without electricity. Everything is made from improvised material: the tank is the glossing machine, the handles are bolts wrapped in plastic, and it sits on two armchair castors. It's supported from the back, but it just lies there. The rest is made up from odd bits of wood. I simply cut a big hole in it for access, so it's easier to pour out the mix. We rolled it all the way from the garage like a steam roller. It's much more efficient to mix with this. It was a shame to throw it away. I paid quite a lot for it originally: 400 Soviet roubles.

Metal drum, bolts, plastic, wood, castors

Yevgeny Petrenko   Kolomna, Moscow region, Russia. 1996

It's just a small hammer. To knock in all sorts of little bits and pieces. Say you need to hammer in a tiny nail. You take it so its head sticks to the magnet, and you hammer it in. You have to think about your fingers, right? My wife does. It's all because of her. Do this, do that! They're all artists, but they can't even hammer a nail in! So I had no choice but come up with this trick, so that my wife wouldn't hammer her fingers off. I drilled a small hole and put a magnet in. Well, really it's more for drawing pins. My wife is an art teacher and she always has to pin up loads of sheets on boards. So I felt sorry for her and so, in a way, I invented a magical hammer. But she still manages to bash her fingers.

Hammer, magnet

Yevgeny Petrenko   Kolomna, Moscow region, Russia. 1997

It's all for our doggy, for her food, so the food goes down smoothly, straight to her stomach. It's supposed to be like that: they say a dog should eat at a certain angle, that's why we made it for her. It hasn't got a name, it's just a little table, a little dog table. The doggy eats once a day. We made it out of what we could find – this little board and other pieces. We didn't give it any special thought. You're the one who's interested in it – we're not interested in it at all! Do you want to keep the dog? We'll give you the table too. Don't you want to have a friend?

Wood, canvas, tacks

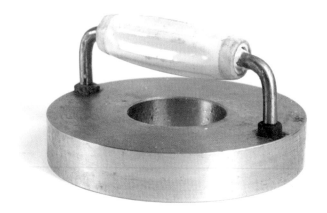

Yevgeny Petrenko   Kolomna, Moscow region, Russia. 2003

It's a press for chargrilled chicken. In Russian it's called a 'collar', from the good old days of stagnation when you could steal metal like this – a stainless steel 'collar'. It's a nice one, isn't it? It's from aluminium piping. The pipes are joined with flanges. Then you screw a handle to it and you get an excellent press for your chicken. I love chicken! The better you press them, the quicker you fry them. The heavier the better. It's all for my family. I don't know why you have to press it down. That's what you're supposed to do, they say. We use it everywhere, to press something down, smash something, or bend it.

Steel collar, handle, bolts

Nikolay Fyodorov   Kaluga region, Russia. 1999

This is a pair of clippers, for metal. The ones in the shops are sold without stands. That's fine when you have a log nearby, so you can rest them on it, but if you don't? Personally, I don't have any spare pieces of wood, but I still constantly need to cut things. I had to give it some thought. It's really practical to prop them up on some kind of stand – then you can hold the material with one hand and cut it with the other. There were some scrap ends of a water-pipe outside, and my neighbour has a welding workshop in his shed. He's a good welder. He quickly put this together for me, just like we'd agreed. We always help each other out, like good neighbours do. You can't survive in the countryside otherwise. One day I help him, the next he helps me. We made these portable, so we don't have to drag the material closer to them. Sometimes I'm the one using them, other times my neighbour might use them. They are clippers 'to share'.

Clippers, water pipe, metal plate

SHOVEL STAND

Sergei Debov   Ivanovo, Russia. 2002

Ah ha! Here's another interesting thing. It was used as a shoe rack. It's the wooden frame from an old TV. They don't make them like this any more, the frame is good, proper lacquered veneer! So it came in handy for shoes, and now for shovels.

We went to our friends, and my friend's wife, Vika, said: 'Seryozha, you're good with your hands, why don't you make me a shovel stand?' She showed me this thing that some lads had made her. I looked at it and said, 'Vika, it's a piece of cake! I'll just take the old box, drill some holes and it will be a stand'. So here you are, I just drilled some holes and it works as a stand.

Wood

Oleg Makarov   Ryazan, Russia. 1980

This was probably made around 1980. Back then I used to go hiking in the mountains and take photographs. There was this special film roll for slides called 'Orvo', it was German, I think. We could get the film but not the frames. We had colour film too, but the quality was much worse colour wise. So you'd keep making your prints, but how were you supposed to show them? Here in Ryazan, we'd sometimes find cardboard slide frames in shops, but they would bend, break, and get dirty. At that time, I was working at a machinery factory during the day, and studying at the Radiotechnical University in the evenings – where we used to hang out at the climbing club. Well, there were computer-controlled machines that we used to program on my shop floor. Basically, we had the need, we had the material, and we had the equipment to hand. So I selected the right milling program, and cut a load of these frames. Then I glued them together. The glue gives it the thickness, and the slide is inserted into the gap.

Hardboard, glue

Alexandr Gerasimovitch Mnatsakanov   Moscow, Russia. 1996

As recounted by his son Georgy: It was my dad who made it. He was born in 1918. There was no sewing-machine for leather so he made each hole and sewed it by hand. There were leather rucksacks in shops, but what's the point in buying one if you can make a more practical one out of a bag that is ten times cheaper? This one has more space, it's bigger than the small rucksacks you buy, and it lasts for ever. My dad enjoyed fiddling around with things like that, and leather was his favourite material. This part's from his WWII military bag. He sacrificed that bag and still put so much work into making this rucksack – about 100 metres of thread! He made some holes at the top here, and put the rope through. The rivets were torn off another bag. At that time it cost next to nothing to have new soles put on your shoes, but he'd still do it himself. He'd buy a piece of leather, cut out the soles, heel the shoes. This was around 1996. This rucksack was used every single day, you could fit all sorts of books in it. The thread was made using a technique all of his own: he'd wax it to make it more durable. This one's actually been sewn together with fishing line.

Leather, rope, fishing line, buckles, rivets

Vassily Nikolaevitch Nikonov   Perm, Russia. 1992

We call it a chopper, for chopping cabbage or anything really. It's a must-have. When you cut with a knife you use your wrist, and it gets tired. And when you chop you use your shoulder. A shoulder is stronger than a wrist – it doesn't get so tired. If you had tried to chop two or three cabbage-heads, you wouldn't be asking these questions. I made it some ten or fifteen years ago, when I was still working at the factory. When they stopped paying us our wages, we instantly asked ourselves the question: how are we supposed to live? It's fine in summertime – the allotment helps out, but what about winter? Then the preserves help us out. So we really need this chopper.

Metal, perspex

FOLDING SCHOOL DESK

Vassily Tyrsyne   Ilyinskaya, Moscow region, Russia. 2007

Our daughter had just started school, so she needed a place to do her homework. There isn't much room – just the table in the kitchen. That's where she'd sit at first, but it wasn't practical. When my wife cooks she needs to use the table, and it's so small that there was no room for homework.

One day I walked past a scrapheap and I saw this old frame from a folding table lying there. I thought I could make something out of it. It was just the right size – a bigger one wouldn't have fitted into the room. I fixed a slab of chipboard for the desk, and made the back of the seat softer, then I oiled the screws so it was easier to fold. It folds and unfolds like this. I'm at work during the day, so I showed my wife how to unfold it. Now everyone's happy.

Metal folding table frame, chipboard, cushion, screws

Yuri Antonovitch Komarov   Kolomna, Russia. 1999

Our boss used to have a book on weaving, so I came up with this idea and made it for myself, for work. When they deliver bricks to our building site, they're bound together using this plastic binding tape. They unload the bricks and just leave the tape. Look how many pieces of tape there are here – 200 lengths! It doesn't break. There's more black tape around now, I've found red too, but not very much. If there'd been more variety, I could have made something better. I'd seen a bloke with a bag like this, so I made this one for myself. I put my thermos in there, some food, whatever I need. A plastic bag? Huh! You can only use it once. You can't find something like this in a shop. I made it narrowish, to fit my things. I wanted to weave it so that if it hits something the bottles won't break.

I retired when this perestroika thing started – there was no work, so I resigned. I stayed at home for about a year, then I found this job working as a handyman.

Plastic binding tape, buckle

Konstantin Ivanovitch Shashkov   Ryazan region, Russia. 1990

My mother liked it a lot when she was still alive. I don't know why. There's nothing particularly good about it. It's probably because when they were dekulakised, that's the only piece of furniture they were able to keep. It got all wobbly, and my mother asked me to fix it somehow.

I'm no carpenter, how could I fix it? There was some wire in the shed, so I tried to pull it together. But either I didn't do it properly, or the wood is too old. It doesn't work, I didn't really fix it. But unless you sit on it, it's actually OK, you can put a cup or a plate on it. It stood by her bed, she wanted it there even though she had normal furniture in the house.

Wood, wire

Viktor Petrovitch Salnikov   Moscow, Russia. 2005

I've always liked sports, I did all sorts when I was young: skiing, running, jumping, volleyball, football, tennis. I used to be a semi-professional in five different sports! But my injury closed all those doors of opportunity. What could I do? I couldn't live without sport. I knew the rules so I went into refereeing. Volleyball, tennis, table-tennis. But there were no whistles then. There were plastic ones for duck hunting and so on, but none with a proper shrill sound for sports. A traffic policeman I knew sold me a whistle – I used it for about five years until I lost it: it was probably stolen. Then I made this one. Luckily, I had everything to hand. I soldered two 50 rouble coins together, as they had no value after the reforms. The whistle came out very nicely. And then, about seven years ago, the guys bought me a real whistle from Germany. It sounds sharper, it's the right kind. If I didn't have the German one, I wouldn't let it out of my sight.

Metal, coins, wood, string

Sergei   Perm, Russia. 2008

We got the cushions where we could, then we just screwed them on. That's all. Fuck knows where it's from, I just put it out and sit on it. That's all. I got those legs from the scrapheap. I put it all together about two or three weeks ago I think. It's just so that you can sit comfortably. You can see everything sitting here. I'm just a janitor and the other guys are guards… And who's got a good life? The one who steals more lives better – that's obvious.

Wood, metal brackets, cushions

Boris Sergeyevitch Ivanov   Moscow, Russia. 1963

It's called a radiogram or a radiophone – it's got a turntable. The design here is all mine, all… Apart from this bit. Everything else is mine. I spent a lot of time working on it – at least 100 hours. That was back in 1963. I used it from 1963 until 1968, when I bought a tape-recorder. I used to take it with me to all the parties, and out into the countryside where we used to have all kinds of fun.

You just plug it in. The speaker is here, you put your record on here, and this is the speed switch. I wanted to make it play at different speeds, but then over time the rubber roller would wear out. By the way, nobody could switch it on without me – I have the switch here. There, now it's ready. Let's put Adriano Celentano on! I was doing my Master's and simply didn't have any money to buy one. There were reel-to-reel recorders, but they were so bulky. The quality of my one wasn't any worse, it was actually better. Plus those ones were made with one pitch while mine had two, and an amplifier. I'd designed the dimensions beforehand, and then I fitted everything in. But it did take me a long time. The frame is from a small suitcase that my father used all through the war. Here, underneath, I screwed wheels onto it. I put it down a lot, you see, carried it through the dirt, on trains… These are the legs, and I made the steel handle too. I used to carry it in my rucksack and on skis. It was very expensive to buy something like this – totally unaffordable for a Master's student. If I wanted, I could even make announcements with it, here's a microphone…

Metal, plastic, rubber, suitcase, castors

Boris Sergeyevitch Ivanov   Moscow, Russia. 2000

One, two, three, four, five… Well, the thing is that I accidentally found the body for it at work. I measured it and realised the size was more or less right. I curved it a little here, and put the plug pins in here. That's it… This is perspex, this is brass, these are copper tubes. I found the piece of perspex with holes drilled all over it, but I still had to drill one or two myself. The pins are fixed here with screws, and the rest was done with a soldering iron. So it becomes a five-way plug. Then, knowing the measurements, I made it so that we could plug it in. You can run five things off it, if they're not too powerful.

Perspex, brass screws, copper pins

CHILD'S MITTENS

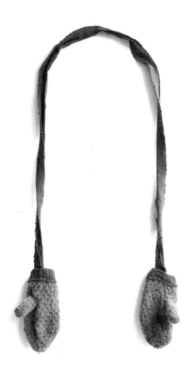

 **Lidiya Sergeyevna Titova**   Kolomna, Russia. 1981

As recounted by her daughter Natasha, pictured: We never had any money –
my parents were paying off their mortgage. I would always be told off for having
holes in my stockings, because you couldn't find any in the shops. I still have
a hat made from rabbit fur that my older sister wore first, then me, and then
my nephew. These mittens were sewn for me by my mum, from Dad's old
jumper. Now you can buy Chinese mittens for twenty roubles. In Soviet
times everyone used to knit children's mittens. An elastic string or a strap
would be sewn to them, so they wouldn't get lost.

Wool, elastic ribbon

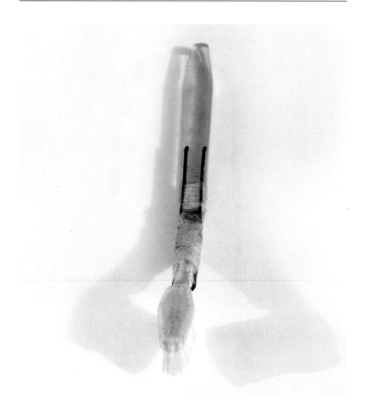

Natalya Saltykova   Moscow, Russia. 2005

I was brushing my teeth when the toothbrush broke, and it was too late to go anywhere – all the shops were already closed. I can't go to bed without brushing my teeth, so I had to come up with something. There were different hair clips and pins lying around in the bathroom. So I took one and fixed it to the handle with tape. There – as simple as that. I used it until I bought a new toothbrush. It's a good repair, it didn't break again.

Hair clips, pins, tape

Pavel Buchkov   Zhukovsky, Moscow region, Russia. 2007

I made it last year I think. I don't care if you can buy doorhandles in the shops
these days. I wanted to have a handle on the door, so I simply took a cutter
and cut it all out. I just approximated the shape of the handle itself. Yes, I did
it all myself – who else would? How funny you are! No, it was my uncle!

Metal

Alyona Vassilyevna Ivanova   Ilyinskaya, Moscow region, Russia. 2008

This is an ice-cream bag, so the ice cream doesn't melt in summer. We hardly ever work on suburban trains in winter – it's cold, and people don't buy that much ice cream. But in summer it's really handy. At least then we manage to make a little bit of money, although it's ridiculous compared to the price of things today. It comes to between five and seven roubles for each ice cream. You can't sell it for more – people don't buy it and it melts. Then you might as well throw the bag away: it'll stink. These bags last about three or four months. The inside is made from cigarette carton cardboard. You cover it with this insulation material – they sell it at the market. Then you wrap it all with tape to make it stronger. Then you attach the handles, and that's it. It's all very simple, and you take as much as you can carry, but you won't manage more than 15 kg, and you don't need to either: you just go to the factory to get more stock. We get it here, in Lyubertsy, because it's convenient and cheap.

Cardboard, insulation, tape

Konstantin Nikiforovitch   Bykovo, Moscow region, Russia. 1977

It's a wheelbarrow. A trolley. What else would you call it? The wheels took me
the longest. My neighbour gave me the tank. It used to be a fuel tank. It all
went fucking rusty. This part was already here, and this one I fucking welded
myself. I used to work as a welder. That's it. Then I could take it over the hills.
I had to go over the hills to reach the meadow to get clay. It hasn't been used
for anything else – that's why it's got rusty. I used to drink wine, I'm an
alcoholic. I thought it would be difficult for me to take clay over the hills, so
I made this thing. I've been retired for twelve years now. A bomb or two fell
on that meadow during the war, there was this big hole in the ground, and
people used to go there to get clay for their ovens. So I went there too. Then
they swamped the whole thing. I thought I'd use this for something else, like
bricks and stuff, but I never used it again.

Metal tubing, metal plate, fuel tank

Ivan Sokolov　　　　Nikola-Lenivets, Kaluga region, Russia. 1998

I made this when I was still living in my old house. It's a sanding tool, well, a kind of file, I guess. But this one is for soft material. You know, if you're sealing a shoe, and you want the glue to stick properly – so that the rubber patch stays – then you have to sand the surface. What can you use to do that? A sanding tool, like this one. It's made from tin, I think, or some kind of galvanized metal. The holes were made with a nail. I made a few holes and bent it onto the stick. Where else could I find something like this, for fuck's sake?

Metal, wood

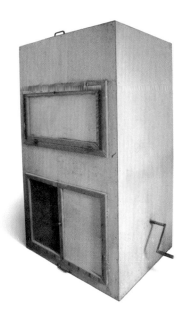

Quico Estivill   Girona, Spain. 1954

It's a resin machine, I made it with a friend of mine about thirty years ago, when I started doing etchings. Artists usually use them. I invented this one myself, I saw something similar and then I added a ventilation system.

Down here, we put some powder, pine resin. We use about 5 kg of well crushed resin. Then we close it. Here we put some pieces of timber, onto which we'll place the sheet. First it spins, and the resin powder rises, like a cloud. Then we open it, and introduce a sheet of copper or zinc or whichever material we want to use. We close it again, and wait a few minutes, and the powder has coated the sheet. It's all about the amount of resin coating on the sheet because that's what's etched into: it brings a uniform tone to the whole sheet, which you can then manipulate through the acid.

I saw this type of machine, or maybe I just saw a drawing in a book, I don't remember where. I saw some description or picture and I thought: 'I know how to do that!'

Wood, metal, hinges

FILM ASSEMBLY AND REWINDING MACHINE

Antoni Vares Martinell   Girona, Spain. 1954

As recounted by his daughter, pictured: Let's see, this is for when the film is finished, to rewind it backwards and forwards. Or to assemble it, to put it back together if it's been broken. Well, this film is actually wider than the reel. This is an 8.5 mm reel. To put the film together – for example if this was broken – then they would scratch the film slightly, cut it and join both parts. It's very basic, because afterwards I am not quite sure how they would press the parts together. Probably just using their fingers. More modern ones have a click that closes and keeps the film together. This should roll, but the film is wider, so I am not quite sure if it will.

My father made it himself. This has been taken from an existing object, who knows what? A drill has this type of wheel, right? This part is from a sewing machine maybe? This could be part of a coffee machine? And of course, for these pieces there should be two buttons since the reels have two holes, one wider than the other. Then they should fit.

I think it's from the fifties, I've seen it all my life, so I am sure that he'd already made it before I was born. He had one before that was older and even more basic. It just had two wooden pieces, and was manual, with no wheel. This one is probably more developed, since there is this handle to wind the film. It was used to rewind the film on the reel, after it had gone through the projector. To rewind it back to the beginning. There was no money at the time, so everything in amateur cinema was handmade.

Wood, metal winder mechanism, spools

CRANE

Joan Ensesa Girona, Spain. 1964

This is the way cranes were made at that time. I made it with my father, back in the sixties. That was when we used it, but we don't any more. Everything was manual, nothing electric. It's made from an iron sheet with some wheels, and this is the arm that has a pulley system hooked onto it, so you can make things go up and down. That's the counterweight, so the crane doesn't topple. It's just a barrel full of sand, nothing else. Simple and rustic. This handle has a lever so you can move it in any direction. We used it to lift large pieces that we had to coat, or figures that needed to be coated, since they were very heavy and we couldn't lift them. We used the crane to pull them up and lift them into the bath, to gold or silver plate the objects. That's its purpose. The pulley system is missing, but I suppose it's somewhere around here.

Sheet iron, wheels, steel pole, oil drum, sand, wire, chain

Joan Fabrega   Girona, Spain. 1998

We use it for the hoist, you just need to hook this here, and then lift it up.
It was made maybe ten or fifteen years ago.

Metal

TOY SPACESHIP

Oriol Lamelas and Josep Ll. Sanchez   Girona, Spain. 2006

Creator's photograph withheld at their request: My grandfather and I built this
spaceship. This part here is a small bottle, I'm not quite sure what it's made
of, but the rest is made of paper and cardboard. I found all the material at home.

Papier mâché, plastic bottle, cardboard

Oriol Lamelas and Josep Ll. Sanchez   Girona, Spain. 2003

I use this string to pull it so that it moves forward, and then I place figures inside. I made it with my grandfather out of cardboard then I coloured it with felt-tip pens and used bottle tops for wheels.

Paper, cardboard, bottle tops, glue

Helene Yousse   Gaüses, Spain. 1993

We bought the house in 1992. There was no furniture, so we had to build it. We found the road sign at a scrap yard just around the corner, it's very convenient. On one side of the yard there were a lot of old enamel signs, and on the other there were several table legs. I was a little sly, I took both without specifying what I wanted to do with them. It was very easy, because the scrap man was in love with me, as I was the only girl going into his yard. I bought all the material for the six tables for about 300 pesetas. And I had almost nothing to do. Fantastic, isn't it?

Road sign, table legs

Helene Yousse   Girona, Spain. 2005

It's like a fountain, it's very nice. The kid turns the tap on, but he can't drink more water than there is inside. Water is vital and you need to take care of it. So the kid has learned from an early age that every time he turns the tap on, it is necessary to turn it off. Otherwise it's a waste, and water can run out.

You can use it to brush your teeth. This is a normal tap with a copper pipe. This part is a nightstand, and we put a small bathroom sink on top of it, and a bucket underneath to collect the water. I'm not sure where it is now. To prevent the kid playing with the water, it was necessary to lock it. It was fantastic because he knew that when it was empty down at the bottom, it was full up at the top.

Tap, plastic bin, copper pipe, bedside table, sink

Maria Gispert Girona, Spain. 2005

Well, this construction is adapted from a clothes hanger. When the girls wanted to dry their hands, it was difficult to use the paper towel because the roll was so heavy. Then, with the help of a man, because I wasn't strong enough, I took the hanger with the roll on it and hung it here. I recycled the hanger to make it easier to dry your hands. You can buy hangers for rolls, but not this size. It was easy to make. I took a hanger, opened it out to make a circle, then closed it again. Then I hung it here. There's no secret. Since we don't have any proper equipment here, I invented this one. Before we used tissue paper or towels. But when we bought this paper, we came up with the idea of this new object. A recycled clothes hanger turned into a paper towel hanger, nothing more. Otherwise, where do you put this paper? When you unroll it, it falls down and it's heavy...

Clothes hanger

Quim Corominas Gispert   Girona, Spain. 1985

You put the paper completely flat and you just place this object on top of it. Or if we've creased the paper, I take this and put it on top to press it. You just need to apply some pressure for a moment, then you can leave it and it stays pressed. Next day, you have completely flat paper, everything is pressed and ready. We use it mainly for bulk papers, but it'll press covers too. Theoretically, these are flat anyway, but with the weight it would all be completely flat, particularly when the glue is still wet. There are some ceramic tiles inside the wrapping. Depending on the weight you need, you can add more or less, and then you cover them with paper and wrap it with tape. I've had it for maybe twenty years. These are my pressing machines.

Ceramic tiles, paper, tape

Marc Ganahl   Wolfhausen, Switzerland. 1982

The queen bee goes in there. You can see it through the acrylic glass, and here it's blocked. There is a gap here where you put honey and sugar in, the queen bee is in there and she can't get through it. Then you go to the colony and remove a wooden board from the hive and put it inside, so that the bees can enter and talk to the queen, feed her, as she can't do it naturally by herself. Then you remove it, and the bees eat their way in through the honey. This takes quite a long time, and by the end they know each other. The queen bee goes through, starts laying eggs, and the colony lives on.

If you simply throw a queen into the colony, they will massacre it because it has the wrong smell. There are different systems, I'd seen something similar and decided to copy it. It's a brilliant system, and it works. The queen never gets killed. She lays eggs, and there will be more bees. Without a queen the colony dies. A working bee lives for between three to six weeks in summer, winter bees live until spring, but a queen will live for several years.

Wood, nails, perspex, card

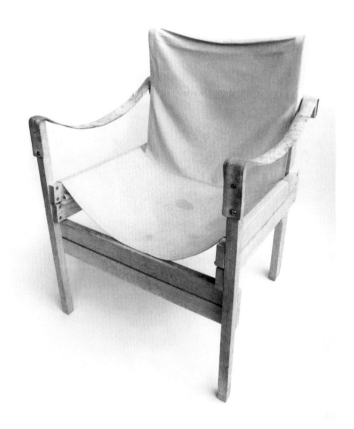

Marc Ganahl Wolfhausen, Switzerland. 1965

It was made more than forty years ago. The wood is bolted together. We'd seen some instructions in a magazine telling you how you could make something like it. Through the company I worked for at the time, I was always visiting a shop where they made wooden models for casting.

I asked them if they could make me something like this – they said yes. I got the wood for free, because they worked a lot for this company and I suppose they thought there would be more work for them.

The fabric used to be much nicer. Back then we didn't have a lot of money, so we never bought chairs from a shop. We had to budget carefully. The cat always sits there.

Wood, fabric, bolts

Willy Vogel   Wolfhausen, Switzerland. 2004

As recounted by Marc Ganahl: The wooden train is made out of old boxes. Willy was about eighty years old when he made it. He had a small workshop at the school and made lots of toys there. He worked as a saddler. He was very skilled. He made all these toys specially for the children: cranes, trucks and so on… He also did handicrafts with the kids. He made them for his grandchildren too. It was what he was interested in, he loved to work with his hands.

Wood, glue, nails

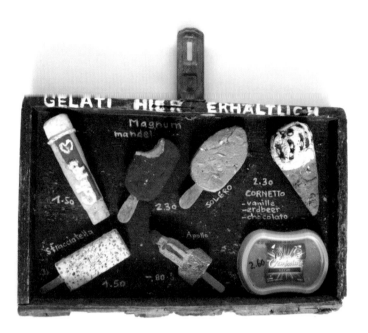

Ursula Gerber   Intragna, Switzerland. 1999

To my mind ordinary ice-cream menus are really boring. I didn't want something like that, I think they're terrible. For me, it has to look authentic, to be appetising. Children always come and look at it. I've bitten into this one myself so it looks genuine. I came up with this idea on the spur of the moment, and it works perfectly. I've wanted to make another one for a very long time, nearly ten years. But I never have the time, never mind...

If you don't have it, you make it: it's the same with everything. This is blue insulating material, very fine, not coarse-grained. I like it because it looks so real, authentic. You can buy all the ice-cream flavours displayed here. The ice-cream is standard though, it's not home-made.

Wood, insulation foam

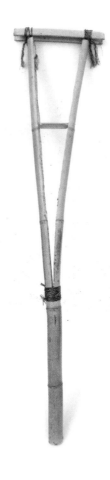

Jonas Herrmann   Intragna, Switzerland. 2009

As recounted by Ursula Gerber: His colleague had twisted his ankle up in the mountains. It happened when he jumped over the wall to get his football. It was hard for him to walk, and so Jonas made him this crutch. He's such a nature man with his long hair, and he's such a special guy – everything has to come from nature – he's a bit of a Robin Hood. The crutches were steady and very solid. The bamboo is from these parts; it grows in the forest. For proper crutches, he would've needed to go down the valley to the hospital.

Bamboo, string

Christiane Kuster   Zurich, Switzerland. 2006

This is a wine box. There were a couple of bottles of wine in there, but I can't remember if I drank them myself. These are special nails. I got them from a roofer. They're carpenter's nails, which are hammered into roof beams. He gave them to me. There were different nails, and I chose the ones that vibrated the best, the ones with the best sound. I didn't put them in the scale order on purpose. I just hung them up according to their size, so that children can experience that the smaller nail gives the highest note. It doesn't have to be a tune. It's a note. It makes a sound. It could have also been made with old keys.

You can't buy one like this anywhere. In a music shop you only get perfect musical instruments, and children have to be careful with them, careful not to break them. With this one they can experiment. It shows the children that you don't always have to throw everything away.

Wood, nails, hook, thread

Luciano Colella   Winterthur, Switzerland. 1980

Wood always gets broken in the garden. So I made the handle of the hammer out of metal. If you use a hammer for a long time, the wooden handle eventually breaks. This hammer's handle is made out of an old metal tube, which I welded. Actually, I made two of them, but one got stolen.

Metal

Luciano Colella   Winterthur, Switzerland. 1980

This is a pickaxe with teeth, which I cut out of metal. In our workshop there is always some scrap metal that no one is using. Anytime I have an idea I can quickly make something. It's been a long time since I made this. I really can't remember how long ago it was.

Metal

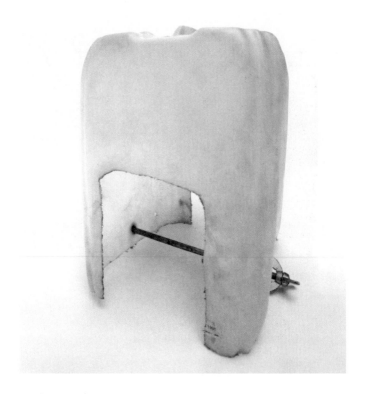

Ferid   Zurich, Switzerland. 2009

This protection 'cap' is roughly the shape of a bell. As you can see, it was made from a plastic carton. I made it six months ago, to protect this machine – the hoist – against wind and snow. At first we covered the hoist with a plastic bag, but it always got blown off. Generally these types of machines aren't protected from the elements. They usually just get rusty and break. But now we've made something. This carton originally contained the cleaning agent for the hoist mechanism. I spotted it and had the idea of making the protection cap.

Plastic carton, metal rod, wing nut

Sergei Nekokoshev   Zurich, Switzerland. 2009

These legs belong to the upper black part of this chair. This was my favourite chair, important to me, cosy to sit on, comfortable. Plus, there's a story that came with it, because the weld seam turned out to be very bad. One fine day it broke into two pieces and I landed on the floor, luckily without getting hurt. I felt that it would be such a shame to lose that chair, so I made this construction from the remaining parts. Now the arm rests are more comfortable and the back is wider. It's well built.

Chair parts

Hannes Schneider Intragna, Switzerland. 2002

The legs of these two chairs were broken, but instead of throwing them away I welded them like this. The problem is... Well, if a couple have had an argument it's OK. But if they have only just fallen in love, they've got a problem. The legs are coated with zinc, and it's almost impossible to weld them. I only managed to do it by fixing the broken ends together, the hooks are needed to balance the pressure.

Chairs, hooks

Hannes Schneider   Intragna, Switzerland. 2009

I made it myself some twenty years ago. At the time, CDs were a new thing. There wasn't a large choice of racks then. Now I don't need it any more. Now I would make it simpler – just a board, because when you have too many CDs, your problem is that the rack is too small.

Aluminium, bolts

Hannes Schneider   Intragna, Switzerland. 2001

On the houses here we have shingles fixed to the wall, they're like little tiles. I built this workbench about eight years ago, to make and repair them, but its design goes way back. In Africa, they do it with their feet, holding on to the part they're working on – here, in Europe, they invented the workbench, so the hands are free for working.

I just built it with blocks, as a bench, you know. Then I fixed the other parts to it. The bench wasn't original either, I found it. I don't know why it was built to this size. A certain amount of space is required so you have enough room to work. Over here you can stand up, here you can sit: it's convenient. Someone in Cavigliano had one like it, I saw it and copied it.

Wood, nails, bolt

Hannes Schneider　　　　　　　　　　Intragna, Switzerland. 2003

This frame belongs in here, it slots together like this. I use the frame with the shelves to transport things – like food and bread – for example, when we go to a celebration on August 1st (Swiss national holiday). I fix the frame with screws so it can't fall out. Then I can carry fresh Swiss bread up the mountain. It's handmade from wood.

　　　　The inner frame is from a thrift shop. The traditional ones are a different shape, the centre of gravity is lower, but I built this one with the weight higher up, so it's easier to carry.

Wood, screws, straps, rope

Heinrich Fanderl   Winterthur, Switzerland. 1980

I hang these small weights onto the plastic covering the compost, so that it
doesn't fly away with the wind. I made them out of old metal pipes, which I'd
found at the scrap metal yard. I sawed the pipes into small pieces, filled
them with concrete and attached a wire sling to them.

Metal tube, concrete, wire

Heinrich Fanderl   Winterthur, Switzerland. 1980

I built this over thirty years ago. All the components are recycled from the rubbish tip: the handle is from an old pram, the wheels presumably from a children's toy. I even found the plastic barrel – it holds fifty litres. The frame underneath is made from waste wood. I found the tap for the barrel in a friend's basement. Over time I've had to replace some of the parts. I built this barrel trolley because the nearest stand pipe is a few metres away from my garden, and I'm not allowed to carry heavy watering cans anymore. Also, the paths on the allotment are very narrow, so I adjusted the width of the trolley to allow me to reach the back of the garden.

Wood, pram handle, wheels, plastic barrel, tap

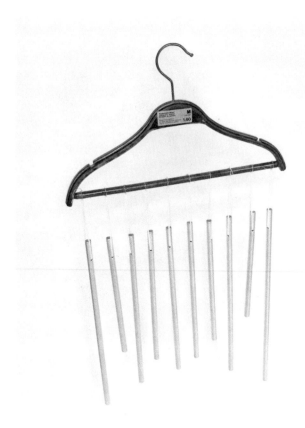

Christiane Kuster   Zurich, Switzerland. 2007

At kindergarten we used to have a children's book. It was a story about an ice cave, and dwarfs who made music with icicles. These pipes sounded just like icicle music to me. I chose the hanger because it makes it easy to hang up, it can be hung anywhere.

Coat hanger, aluminium tubes, thread

Dimasi Giuseppe   Winterthur, Switzerland. 1989

I made this barbell about ten years ago. I'd just built something in the garden, and there was some concrete left over, so I used it for this. The bar used to be an old water pipe, and the concrete is wrapped in sheet metal to give it shape. It's total weight is about 20 kg. When I was at my best, I was able to lift it twenty or thirty times, but I don't train with it any more. If you want, you can take it with you, I don't need it any more.

Concrete, water pipe, sheet metal

 **Adrian Stumm**   Zurich, Switzerland. 2009

This is made from a wine box, I put it together about two months ago. The standard breadboxes aren't very nice, they always fall apart, and then the cutting board doesn't fit in. I don't know how I got the idea, but it's really simple.

Wooden wine crate, nails

Adrian Stumm   Zurich, Switzerland. 2009

This is a coffee grinder. I can take it anywhere. This part is from a bicycle, the stem, where the handlebar goes through. The handle always goes stiff, which makes it exhausting to hold, using this makes it a bit easier. In the previous flat it was attached to a cupboard, but here they're all made of metal, so that won't work. This is my grinder from New Zealand. It's not really that good, it doesn't work very well, because you need a lot of strength. The electric grinder is just better and nicer to use, because you can feel the coffee beans.

Bicycle stem, coffee grinder

Caryl Zeni   Zurich, Switzerland. 2007

I invented this contraption to remove nails from battens. You place the batten across the gap between the pieces of wood, then hit the nails with the hammer. But not all of them at the same time – one after the other. This makes it easy to get the nails out. It works. I developed it a couple of years ago, when I started to work at the sailing club. My system is both safe and useful.

Wood, nails

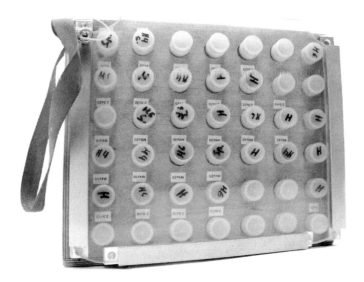

Stefan Schrank   Zurich, Switzerland. 2004

As recounted by Caryl Zeni: This box was made by Stefan Schrank, he was a super tinker. It's specifically for clout nails and screws. If you go out to fix a boat all your pins are kept in their correct places using this box. Normally the clout nails and screws go all over the place, because of the rolling and lurching. There is also the possibility that they might hit and hurt people, so the box helps avoid any injuries. It's safe and simple. On board everything has its own place in self-made boxes, but I'd never seen anything for the clout nails or screws before. It's unique. The containers were made out of test tubes from a chemistry lab. The concept, the idea, and how you use it are all original – designed by the same man. He was a special person, a juggler.

Wood, test tubes, perspex

SPOON

Ivan Kuzmitch Satchivko   Kiev, Ukraine. 1946–1948

As recounted by his son Fyodor Ivanovitch, pictured: I inherited this spoon from
my father. He died not so long ago, and you could say that it's a kind of
memory of him. Here's the story: Kiev was under Nazi occupation – there
wasn't much left of the city, and my father came here to help rebuild it. He'd
been living out in the country until then, but their village was burned down…
Times were hard after the war, people didn't have anything – they lived in huts.
It was here, in Kiev, that my father met my mother. And then they had me.
My father worked in a factory, where he did reconstruction work. When any
spare metal came along, he made things with it for family and friends: a plough
of some sort or a spade – people didn't have anything, you see. Yet there was
a lot of abandoned, broken military equipment left in the fields. One day some
relatives from the country brought him some aluminium wreckage from a
downed German bomber, and asked him to make useful household things out
of it: combs, spoons, mugs, bowls. Nothing went to waste, nothing was left
of the wreckage when he'd finished. We used the crooked saucepans and frying
pans that my father had made for a long time. Then, when our living conditions
improved, we threw everything away. We've still got a couple of these spoons
though. While my father was alive, it never occurred to me to ask him about
the spoons. It's only since he passed away that my mother told all this.

Aluminium

CHISEL SHARPENER

Vassily Perevalsky Kiev, Ukraine. 1965

As recounted by his son Yevdokim, pictured: There, it's a special tool to sharpen the – what do you call them? Engraving chisels! For chiselling wood. There's this hole here. You put it in and get the angle you need for your sharpening. You press it with your finger here, so that nothing moves. Then you slowly take a small metal bar and begin to sharpen it. Sometimes you need to put a few drops of oil on it so that the chisel doesn't overheat. My father made this one. He used it, and now I do too – it's very practical. You can't find anything like this in shops because everyone uses all sorts of mechanical grindstones to sharpen their chisels now.

Metal plate, wood

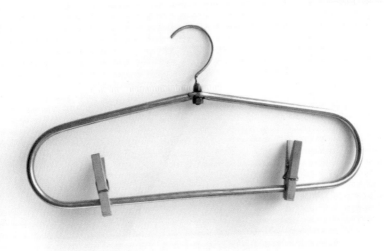

Nina Vasilyevna Medvedeva   Dnepropetrovsk, Ukraine. 1989

As recounted by her husband Viktor: My wife made it about twenty years ago, she's really good with her hands. Now there are plenty of hangers in the shops, but there weren't at that time. Anyway, what would she buy a hanger for if she could make one herself? She made it to hang her clothes up at work. We did have hangers in the house, but why would she take stuff from home to work? Now, the other way around – home from work – that's the right thing to do!

　　　　She'd worked as an assembler all her life until Gorbachev started with his perestroika, until the Soviet Union fell apart. They started with such a mess that it makes me sick even thinking about it. Some days there was work, other days there wasn't. They wouldn't pay us wages for six months, so she went to work as a welder, at least she could earn some money in the welding section, even if it's not assemblage.

Titanium tubing

Arwel Davies   Lloc, Wales. 2000

It's a car jack, it's used for raising the car. I made it three years ago. But it's easier using a proper jack – you wind the handle and then…

Wheels, metal pole

David Catherine   Rhuallt, Wales. 1995

We wheel it out to the garden. If it's in the wrong place we can just move it, you know, turn it around because it's portable. I've even taken it to people's houses to do barbeques. These shelves were taken from a beer fridge in a pub. This used to be an anti-freeze barrel. I got the tray in the middle out of a skip. The wheelbarrow is just an old wheelbarrow. It didn't cost anything.

Wheelbarrow, laminated chipboard, metal barrel, metal grill, wood

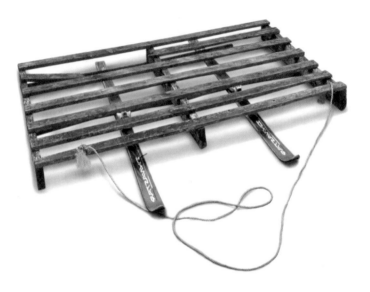

Graham Crossley   Llandudno, Wales. 1998

We use this to move mats across the ski slope, so they slide on the surface of the slope, because they're quite heavy to carry around. We constantly have to repair the slope, so we need to move the bits of matting around quite often. We sometimes get farmers to take the matting away. Since we've been here we've gone through about ten of them. It's just a sledge really.

Pallet, skis, rope

Graham Crossley   Llandudno, Wales. 2003

Basically, one of them has the time out, which is 1.30 pm, and the other one is 3 o'clock, which is the time you come in: because you get an hour and a half skiing. It gives people an indication. When skiers come here to ski, the normal session is one and a half hours. So they need to know what time they have to come in. This is the time now, this is the time they are going out to ski, which means that this is the time they will come back in. We are busy in the summer because we have limited equipment, which limits the number of people on the slope. On the receipts we write the time, which makes life easier for our staff, so they don't have to work it out. Each time somebody comes, they just glance up there and write it out on the receipt.

Clocks, stickers

David Marshall   Penmachno, Wales. 2001

It's called an outrigger. The canoe is unstable and in the Pacific they used to make an outrigger from bamboo and wood, which makes the canoe very stable. So I made this one from plastic. Those are drain pipes, and on the end I've a painted wooden cone. I made the boat too. I don't buy things if I can make them better.

Plastic drain pipe, wood

Craig Williams   Llandudno, Wales. 1997

This one here is to take the wheels off carriages. It's based on a wheel bearing extraction device – so the design has been pinched. You see, once you've removed this nut from here, you then tighten this on the poles of the wheel, and then pull the wheel away. It's a very simple device. I just strengthened it by welding these pieces there, to stop it from bending outwards. There is actually a tool specially made for this job. We're just lazy basically, so we built it here. I just sort of made it myself, about four years ago, it hasn't got a name.

Metal

David Marshall   Penmachno, Wales. 1997

I make two tracks for the wheels of the car. I walk on behind and push the snow off the road. I push this and it pushes the snow. A snowplough on the road, on a lorry, only has one side, so the snow goes on the side of the road. I made this with two sides.

Wood, nails

Alan H. Morley   Conwy, Wales. 2000

With this object, well, we wanted to sell honey bees, queen bees. So we must've made about fifty of these in the workshop during the winter months. You put the queen bee in here.

Metal sheet, wire mesh, wood

SACK BARROW

Freddie Armour-Brown   Maerdy, Wales. 1991

This is no ordinary sack barrow, I can lift a really large rock with this barrow, like that one there. If you're doing building work and you want to move a rock to put it in the wall, this is the tool. You can't buy very strong ones, this one is strong enough to carry a huge rock. This piece of metal is the spring from a lorry. Very strong. I've never seen them this size.

Metal, wheels

Alun Jones   Trofarth, Wales. 2000

We cut the containers. We cut holes there for the sheep to get the hay. It's an empty five-gallon drum. Well, twenty-five litres. It's a feeder. I use an electric grinder to make the holes, then it's done.

Plastic container

Eifion Jones   Trofarth, Wales. 1998

As recounted by his father Alun: Who made it? The son! The son! Well, I suppose it was wooden before, a few years ago, wasn't it. It was a wooden handle, and it was breaking all the time. And, he welded it, with the electric welder. It was a while back. Well, you know, gosh, six to ten years or so… My son Eifion, he's the welder. We've got an electric welder, that's the welder there. It's pretty handy to have a welder on the farm.

Metal rod, hammer head

Simon Greaves   Llandrillo, Wales. 2001

I have a ten-year-old who nagged me to make him a hard suit of armour, as if it was the easiest thing in the world to do. It's actually interesting because we found out the other day that our surname – Greaves – means armourer. Specifically, a 'greave' is that part of the armour that protects the knees. I made this a year ago from some scrap copper that had been stripped from the roof. Unfortunately it isn't very comfortable.

Copper sheeting

Simon Greaves   Llandrillo, Wales. 2001

It's a go-kart for my son to ride down the hill. He wanted to make one so I helped him, and we used whatever materials we could find. We had bought him one, but he wasn't satisfied with that. I just made it from pieces lying around. I don't actually know where these bits came from. I think this was some kind of shopping trolley or something, I'm not sure. The wheels were from two different scrapped carts. My father-in-law had them in his shed. He doesn't throw anything away. He's a good man to know if you want bits like this.

Trolly parts, wood, cushioned seat, rope

Simon Greaves Llandrillo, Wales. 1998

When the slates go on the roof they have to overlap by a certain amount, so the hole has to be gauged just right so the slate below is covered: so there's no water penetration. Then the next slate on the roof would be in the same position as this, it would come down to here, so that the water runs over the center of the slate below. That's all this device does. I use it to set the same distance that I need. Then I tip a hole in on this line here, and the hammer just knocks a hole through. And then you nail it on the roof. This is iron, just a piece I found. The iron supports the slate. If you hit it on something soft, it bounces, it doesn't go through.

Wood, iron bar, nails

Hylda Morley   Conwy, Wales. 1975

Hot-water bottle, light fitting, shade

Tim Hilfe   Llanfairfechan, Wales. 2000

That one there? Yeah, it's just a tyre with a bucket in it. That's a Land Rover tyre. I've made it like that so the horses can't pick it up and turn it over. If you tie this one up, she'll undo the knot and go free. Then she'll undo everybody's knots. She could turn over tons of them. But that's absolutely typical of most Welsh Ponies.

Tyre, plastic bucket, rope

HORSE FEEDER

Tim Hilfe   Llanfairfechan, Wales. 1998

I made that, yes. It must have been about two or three years ago. The horses chew it, that's why the edges are like that. You can buy metal versions, but I made it out of scrap. I'm always looking for things that nobody else needs. This part is to catch the hay that falls through. When you put the hay in, it drops through when they're eating it. It falls through the holes and it lands on the board, and they can eat it up from there. It doesn't go on the ground then. All the wooden bits I made, the rest of it I bought or was given.

Wood, metal fencing

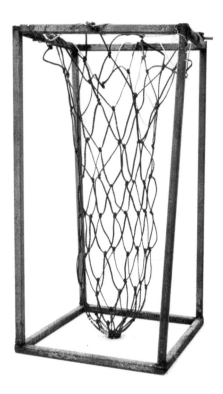

 **Michael Humphries**   Llanfairfechan, Wales. 1999

This thing I'm going to show you now was made by a friend of my son about four or five years ago. He was standing here and he said, 'I know what you need', and he made this, and it works! It's a... I had to repair it this morning. It's been repaired and used, and repaired and used, and repaired. But it's a... It just works. It's dead simple. We put hay in it for the ponies. A friend of my son made it for me. Dead easy. He went off and made it, literally just cut up some metal and made it. I always thought it was a brilliant idea, it was so simple.

Metal, netting

Phil Morris   Groesffordd Marli, Wales. 2000

These are markers for mole trapping. There's three parts to one thing. This is a marker to find the trap. It also stops the wild animals taking it out because it's captured there, at the bottom. You see? This is a hazel peg, which is again, just lying around. We use that material because it's a good quality seal. This is blue, and it blows in the wind so you can find it. So the trap's set for a mole in the ground, and that's pegged to mark it. And when you're picking up a lot of traps during the daytime you don't have to bend over every time.

Hazel branch, nylon string, tape

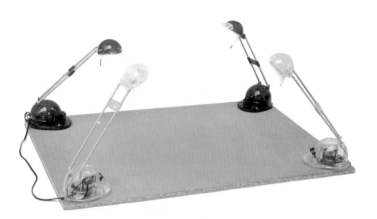

Tim Williams   Pontllyfni, Wales. 2003

It's for copying artwork onto slides. You place the photo in the middle of the board and switch the lights on, so you've got even lighting. Then you photograph it from the top. Normally when you buy these they cost a lot of money. But I got these lights for fifteen quid in an Ikea sale, and a piece of cheap wood. You'll never look at my slides the same way again...

Plywood, table lamps

CYCLING OVERSHOES

Tim Williams   Pontllyfni, Wales. 1994

Well, I was with my friend on a cycling trip to Norway, and it didn't stop raining
for the first three weeks. I didn't have any overshoes. So basically, my feet were
soaked. It was so bad that in the end, after three weeks, they started to
develop a kind of trench foot. When you've been in the bath for so long your
feet go like prunes – well it was just like that. Continuously. So I decided to
make some overshoes out of some Norwegian fishing mitts. You use this cord,
you know, you just tighten them up at the back. But I spent a whole day making
them, actually sewing up all the eyelets, to make them strong enough. They
were quite effective, they worked well. But the guy I was with, he thought they
looked like birds' beaks. He was so disgusted with these things, that when we
would arrive in a village or town he asked me to take them off. I cut the thumb
off, and split it down the centre, because otherwise there's no way you can get
your foot into the mitt. If you're on a pushbike, all the spray comes from the
front wheel onto the top of your foot. So this is the area you need to protect.

Fishing mitts, cord

INDEX

The author would like to thank the following people for their collaboration, financial and moral support:

MXM-gallery, National gallery, Praha, and personally Marin Zet, Tomáš Glanc; Ikon Gallery, Birmingham, and personally Jonathan Watkins, Debbie Kermode, Julia Obissova-Fernee; Shrewsbury Museum & Art Gallery and personally Adrian Plant; Oriel Mostyn Gallery, Llandudno, and personally Martin Barlow, Anders Pleass; Kunsthalle, Baden-Baden, and personally Matthias Winzen, Nicole Fritz, Victor Agafonov; Kunstverein, Rosenheim, and personally Iris Truebswetter; Alte Kelter hall, Fellbach, and personally Marina Rantzau; Festival der Regionen, and personally Martin Fritz, Susanna Posegga and Paul Vallen; Liga-gallery, Kolomna, and personally Vitaliy Khitrov; Tirana Biennale of Art, and personally Edi Muka, Ada Kapaj, Prizreni Ingrit; Burren College of Art, Ballyvaughan, Co.Clare, and personally Fiona Woods, Siobhan Mulcahy, Viktor Tcvetkov, Pat and Maria Finucane; Centro per l'arte contemporanea Luigi Pecci, Prato, and personally Stefano Pezzato; Galleria Nina Lumer, Milan, and personally Nina Lumer, Ludovika, Fabrizio Fenghi; Perm Museum of Contemporary Art, Perm, and personally Marat Guelman, Mikhail Surkov; The CAPC – Museum of Modern Art, Bordeaux, and personally Emilie Renard, Viktoria Vinogradova; Natasha Akhmerova Gallery, Zurich, and personally Natasha Akhmerova, Mario Lüscher, Natasha and Daniel Ganahl; Bòlit – Centre d'Art Contemporani de Girona, and personally Rosa Pera Roca, Evdokim Perevalskiy.

And the following people in Russia:

Aleksandr and Vladimir Atkishkin, Andrei Arkhipov, Mikhail Akhmatov, Dmitrii Bragin, Nikolai Brizgov, Aleksander Brodskii, Maria Chekmenyova, Marina Chizhkova, Andrey Drozhdov, Vladimir Dukelskii, Irina Fadeeva, Bart Goldhorn, Vladimir Krichevskii, Vladimir Kupreyanov, Andrei Lazarev, Valentina Lopatina, Kirill Medvedev, Viktor Miziano, Georgii Mnatsakanov, Yuri Nikich, Aleksei Orekhov, Valeriy Podoroga, Vera Pogodina, Nikolai Poliskii, Olga Potapova, Andrei Repin, Valerii Rodin, Natalaya Saltykova, David Sarkisyan, Asya Silaeva, Sergey Sitar, Vladislav Sofronov, Marina Starush, Aleksandr Tarasov, Vasilij Tsereteli, Lidiya Vasileva, Julia Vainzof, Irina Yurna.

Vladimir Arkhipov

Contemporary material culture keeps slipping away from attempts to describe it. Yet I hope that in this book we have managed, to some degree, to define the European traits of the universal phenomenon of home-made practical objects. Just like our own real lives, these real-life objects resist classification and systemisation. For this reason I haven't categorised anything, or kept statistics. I simply try to document every home-made object that I find, in its original context, using my camera and a Dictaphone.

Here is the paradox: there might not seem to be anything so important in the way John McNamara (Ireland) filters his water, or in what the dog belonging to Mario Pieraccini (Italy) sits on, or in what kind of kettle Andrei Repin (Russia) makes his tea. These people, like thousands of others, have little or nothing to do with Art. But they have constructed fascinating forms for these purposes, worthy of exhibition in a museum.

Specialists have created much of our visual environment. They also create its aesthetic character. If there were suddenly no more professional designers or object-makers left in the world, then the process of creating new designs, new forms, would of course not diminish.

According to Aristotle, Art is an imitation of reality, so couldn't creating the reality preceding this imitation itself be an important earlier stage of the artistic process? How is this creativity subtly different from Art?

Our world demands the constant production of images. We have specific people to create them (such as illustrators and designers). They understand the rules of what should comply with what. They are sophisticated and experienced: the results of their efforts are professional. But where does Wonder enter into this equation? How can one feed this insatiable demand and still satisfy the thirst for revelation?

Art too is a long way from exhausting its democratic, or any other, potential. It can tend to follow the business world, failing profound needs. We suffer from selfishness and formality. If Art wants to avoid turning into design, it has to remember to 'make the moment linger' [Goethe's Faust].

The objects in this book are authentic, and belong to ordinary people; they are evidence of necessary creativity; they are of that special class of functional objects in the world that were not made to be sold; they have all been authored, and retain the signatures of their authors; they are all unique: and they bear something of the character of works of Art.

TEXT AND PHOTOGRAPHY Vladimir Arkhipov

DESIGN AND EDIT Murray & Sorrell FUEL
CO-ORDINATOR Julia Goumen

The publishers would like to thank the following people for their help
with the translation:

Margarita Baskakova of Fitzrovia Languages
Karin Schranz-Kuhn
Dagmar Braeuchi Schubert
Anne-Caroline Meyer
Nuria Jorba Arimany

Also thanks to Jeremy Deller, Fergal Stapleton, Rebecca Warren.

First published in 2012

Murray & Sorrell FUEL ©
Design & Publishing
33 Fournier Street
London E1 6QE

fuel-design.com

Printed in China

Distributed by Thames & Hudson / D.A.P.
ISBN 978-0-9568962-3-0